MAGNETIC RESONANCE IMAGING

Primers in Biomedical Imaging Devices and Systems Series

MAGNETIC RESONANCE IMAGING

Recording, Reconstruction and Assessment

NILANJAN DEY

Department of Computer Science and Engineering, JIS University, Kolkata, India

V. RAJINIKANTH

Department of Electronics and Instrumentation Engineering, St. Joseph's College of Engineering, Chennai, Tamilnadu, India

ACADEMIC PRESS

An imprint of Elsevier

ELSEVIER

Academic Press is an imprint of Elsevier
125 London Wall, London EC2Y 5AS, United Kingdom
525 B Street, Suite 1650, San Diego, CA 92101, United States
50 Hampshire Street, 5th Floor, Cambridge, MA 02139, United States
The Boulevard, Langford Lane, Kidlington, Oxford OX5 1GB, United Kingdom

Notices

Knowledge and best practice in this field are constantly changing. As new research and experience broaden our understanding, changes in research methods, professional practices, or medical treatment may become necessary.

Practitioners and researchers must always rely on their own experience and knowledge in evaluating and using any information, methods, compounds, or experiments described herein. In using such information or methods they should be mindful of their own safety and the safety of others, including parties for whom they have a professional responsibility.

To the fullest extent of the law, neither the Publisher nor the authors, contributors, or editors, assume any liability for any injury and/or damage to persons or property as a matter of products liability, negligence or otherwise, or from any use or operation of any methods, products, instructions, or ideas contained in the material herein.

Library of Congress Cataloging-in-Publication Data
A catalog record for this book is available from the Library of Congress

British Library Cataloguing-in-Publication Data
A catalogue record for this book is available from the British Library

ISBN: 978-0-12-823401-3

For information on all Academic Press publications visit our website at
https://www.elsevier.com/books-and-journals

Publisher: Mara Conner
Acquisitions Editor: Tim Pitts
Editorial Project Manager: Isabella C. Silva
Production Project Manager: Anitha Sivaraj
Cover Designer: Matthew Limbert

Typeset by TNQ Technologies

Contents

Preface

Biomedical image-supported disease diagnosis is a commonly adopted clinical procedure that is widely used to detect disease in internal and external body organs. Although a considerable number of imaging procedures have been developed and implemented, magnetic resonance imaging (MRI) is widely adopted to diagnose acute diseases in various body organs. The added advantages of MRI compared to other imaging methods are as follows: (1) it is a radiology-supported technique and can be performed with or without a contrast agent; (2) it helps to provide a reconstructed 3D image; and (3) it consists of various modalities, such as Flair, T1, T1C, T2, and DW. The visibility of the disease in MRI is better than in other existing imaging methods.

Recently, MRI-supported disease discovery and treatment implementation have helped to advance the medical industries, and a significant amount of traditional and artificial intelligence technique-based image evaluation methods have been proposed and implemented by the researchers. The MRI of a chosen organ can be evaluated in 3D or 2D form, with the examination by 2D being relatively simpler and hence it is used most often, involving: 3D to 2D conversion, preprocessing, post-processing, and disease detection with the chosen technique. Further, 2D MRI helps to obtain various image planes, such as axial, coronal, and sagittal, which further helps to obtain complete information regarding the disease being detected.

In this book, the examination of 2D MRI slices is discussed with appropriate examples. The execution of the traditional and heuristic algorithm-supported MRI assessment scheme is also presented for segmentation and classification tasks. The experimental investigation presented in this book is demonstrated using the results obtained with MATLAB software.

The book is organized as follows:

Chapter 1 discusses the image-assisted disease-screening procedures normally employed in hospitals to detect disease in various body organs. This chapter discusses the commonly considered bio-signals and bio-images to detect disease, and the merits of bio-images over bio-signals. It also demonstrates the methodologies considered in collecting biomedical data from the patients, their evaluation procedures, and the merits of medical images compared to bio-signals during real-time disease detection in hospitals.

Chapter 2 presents an overview of the MRI recording and reconstruction procedures followed in radiological centers. The chapter also presents details about the MRI recording system that must be considered to collect the images from the patient with and without use of a contrast agent. Further, the visibility of the MRI slices in modalities, such as Flair, T1, T1C, and T2 are discussed with appropriate sample images.

Chapter 3 demonstrates the choice of appropriate MRI examination methods, such as image conversion, preprocessing procedure to enhance an abnormal section, segmentation procedure to extract the section, and its evaluation techniques. In this chapter, the necessary MRI slices collected from benchmark datasets are considered for the assessment, and the results of various enhancement and extraction procedures are presented and discussed.

Chapter 4 presents the examination procedure employed to detect breast tumor using breast MRI slices with planes, such as axial, coronal, and sagittal. The proposed work employs trilevel thresholding with Kapur's entropy for breast tumor enhancement and watershed algorithm-supported segmentation. The performance of the proposed scheme is demonstrated using the recently developed heuristic approach called the Mayfly Algorithm and the eminence of the proposed scheme is confirmed by computing the necessary image quality measures obtained by comparing the segmented tumor with the ground truth.

Chapter 5 demonstrates the examination of heart MRI slices using a chosen image segmentation technique. The task considered in this work is to extract and evaluate an abnormal heart section and is achieved using the heuristic algorithm named the Spotted-Hyena-Algorithm. This work implements a trilevel thresholding, and level-set segmentation procedure to extract the abnormal region and the mined section is then compared with the ground truth to validate the performance of the proposed scheme. This chapter also presents a brief comparison between the images enhanced with Kapur/Otsu function.

Chapter 6 presents an overview of deep-learning supported abnormal region segmentation from the chosen MRI slice. In this chapter, the pretrained models, such as VGG-UNet and VGG-SegNet, are employed to extract and evaluate the tumor section from the brain MRI of modality T2 and the results are presented and discussed. The proposed deep-learning scheme is executed using MATLAB software and the results attained with this approach are compared and verified with the results of a watershed algorithm.

Chapter 7 demonstrates the experimental outcome for the detection of an ischemic stroke lesion from brain MRI slices. In this chapter, the pretrained VGG16 is employed to extract and evaluate the stroke lesion from the MRI of a chosen modality and the complete results obtained during the experimental investigation are presented and discussed. The performance of the proposed scheme is also tested and validated using the brain tumor MRI database.

Dr. Venkatesan Rajinikanth
St. Joseph's College of Engineering

Dr. Nilanjan Dey
JIS University

Introduction to image-assisted disease screening

1.1 Introduction

Recent progress in science and technology has helped the industrial sector to achieve significant progress with innovative and improved products. Because of this, a considerable number of innovative and modern diagnostic procedures have been developed and produced to support the medical industry, helping with (1) quick and accurate disease detection, (2) faster decision-making and treatment implementation, and (3) enhanced recovery rates with life-support devices [1–5]. Modern progress in the medical domain, such as the Internet of Medical Things (IOMT) [6,7], cloud and fog computing [8,9], big data analytics [10–12] and Medical body area network (MBAN) has also extended support toward remote patient monitoring and disease diagnosis [13–15]. Further, the concept of Healthcare 4.0 helped to create an improved healthcare ambiance by including all the necessary innovations to support better healthcare management.

The current growth in the healthcare domain has facilitated rapid growth of state-of-the-art medical amenities. Further, the discovery of a significant number of vaccinations has helped to reduce a number of acute and transmissible diseases. The modern diagnostic amenities existing in multispecialty healthcare centers and scheduled health check-up facilities have also facilitated the detection of diseases in their early phases [16,17].

Although, significant measures have been adopted to avoid and control diseases, the rapid occurrence of new types of acute and infectious diseases has been increasing due to various causes, such as heredity, aging, and other unavoidable causes [18–20]. To support the premature discovery and treatment of these illness, considerable investigative approaches have been developed and used in healthcare clinics [21–23].

External organ diseases can be easily identified by doctors by the symptoms, visual examination, and computer-supported diagnosis. For example, an abnormal cell growth, wound, or infection

Magnetic Resonance Imaging: Recording, Reconstruction and Assessment. https://doi.org/10.1016/B978-0-12-823401-3.00001-8

in an external sensory organ, such as the skin, is easily identified by a doctor through a visual check, and it usually can be easily treated with the recommended medicines. Cell growths, wounds, or infections in internal body organs are more acute and left untreated these often lead to death. Hence, a recommended clinical protocol must be followed for the detection of internal organ abnormalities. Diseases of the internal organs must be diagnosed using recommended procedures, including (1) signal-assisted schemes (SAS) and (2) image-assisted schemes (IAS).

SAS is a commonly employed preliminary screening procedure in which a chosen electrode set-up can be used to record and examine the bioelectric potential developed at the cell or organ level. The recorded signal using a chosen scheme is very complex in nature and needs preprocessing (filter) to remove the unwanted high/low-frequency components which are associated with the actual signal. Further, domain proficiency is essential to evaluate the signal to identify any abnormalities. An electrocardiogram (ECG) is one of the most common biosignals recorded and evaluated to identify abnormalities in the heart [24,25]. The assessment of the ECG can be normally performed by an experienced doctor and/or a dedicated computer algorithm, which is designed to examine the ECG pattern [26,27]. Like the ECG, an electroencephalogram (EEG) is also a well-known biosignal widely used to record brain activity. However, compared to an ECG, patterns in EEGs are quite complex and hence, signal to image conversion methods are widely adopted in the literature to examine brain abnormalities [28,29].

The existing literature in the healthcare field confirms that the vital information seen in bioimages is greater than that for biosignals and bioimages recorded with a chosen modality that can be easily examined by disease experts. If the number of images is less, then a visual examination may be sufficient to identify the cause, type, location, and severity of the disease. Further, a chosen image-processing scheme can be used as the assisting medium to examine the image (gray/RGB scale) and the outcome attained with the assisting medium combined with the suggestion by the doctor helps to identify the most appropriate treatment to control/cure the disease in the internal organ.

In most cases, noninvasive biosignal collection is adopted as the initial practice to screen the internal organ under study. After assessing the signal, an image-supported examination is suggested by the doctor to confirm the disease and its severity. In recent years, modern imaging procedures have been accessible in hospitals to provide essential details about the internal organ under study. Further, the current imaging procedures are able to record digital images, which are easier to store than the earlier imaging schemes.

Medical images can be recorded using a range of procedures, such as a simple digital camera-supported scheme or a radiology-assisted scheme. In bioimages, there is a considerable number of imaging modalities, such as radiography (X-ray), computed tomography (CT), thermal imaging, ultrasound images, magnetic resonance imaging (MRI), magnetic resonance angiography (MRA), etc. [30–35]. However, the SAS still follow the traditional methods and require intricate diagnostic methods compared to IAS. The choice of imaging scheme depends on the internal organ under study and the skill of the physician.

This chapter presents a summary of disease-viewing practices for selected diseases in organs including the breasts, heart, and brain, and the clinical procedures followed to examine these diseases also presented with the appropriate results. Further, various imaging modalities have been used to record and examine abnormalities are also discussed and the suitability of MRI compared with the other imaging modalities is discussed with an appropriate example. This chapter also discusses the information regarding the need for and development of artificial intelligence (AI)-supported abnormality examination procedures to examine MRI results with better accuracy and the validation procedure to confirm its clinical significance.

1.2 Acute diseases in vital organs

Diseases, such as infections, wounds, and other abnormalities in external body organs are easy to diagnose using visual and traditional procedures. Further, the treatment of external body organs is quite easy compared to diseases in internal organs. Most of the internal organs are vital to human physiology and any abnormality in these organs can severely affect the whole physiology, with untreated abnormalities sometime leading to death or permanent illness/disability.

The internal organs, such as the heart, lungs, kidneys, gastric tract, and brain play essential roles in physiology and abnormalities in these organs can severely affect the whole body. Hence, great care must be taken to maintain these organs in a healthy condition. This subsection discusses common diseases in organs, such as the breasts, heart, and brain and their examination methodologies.

1.2.1 Breast abnormality detection

Abnormalities in the breasts are mainly due to aging and other related issues. In women, breast cancer (BC) has emerged as a major life-threatening disease, and its early detection and treatment help to cure it using the recommended treatment procedures, such as radiotherapy, chemotherapy, and surgery. BC is influenced by various factors, including aging. The early form of BC most often detected is ductal carcinoma in situ (DCIS) and the detection procedure involves various phases, including (1) visual inspection by an experienced doctor, (2) image-assisted detection, and (3) needle biopsy-based confirmation [36,37].

When a patient is diagnosed with BC, the stage and severity of the cancer need to be identified using various imaging procedures ranging from the organ level to tissue/cell-level imaging. Organ-level diagnosis can be achieved using imaging modalities, such as mammography, thermal imaging, ultrasound, and breast MRI, and tissue/cell-level assessment can be done using microscopy images obtained from a needle biopsy sample [38−41]. Needle biopsy is a painful invasive procedure widely used to confirm the stage of BC, and the initial diagnosis can be performed using common noninvasive imaging procedures including radiology-supported imaging (MRI).

Fig. 1.1 shows the various stages involved in the BC detection process. BC is one of the most severe illnesses in women and the incidence rate is increasing globally [42]. The modern diagnostic procedures available in the hospitals identify and treat this disease with scheduled screening irrespective of the patient's health condition. When BC symptoms are identified during a self-check and/or visual examination by a doctor, image-supported detection is recommended. Based on the doctor's suggestion, the suspicious breast section is recorded using a chosen image modality and this image is assessed using a computerized algorithm. In some severe cases, the breast tissue sample is collected using a needle biopsy and the collected sample is further investigated using the images obtained from a digital microscope.

The early form/developed BC is mainly due to abnormal growth inside the breast duct and surrounding tissue. In practice, mammography and breast MRI are the most widely adopted image modalities due to their reputation and success rate in detecting abnormalities. Along with above procedures, other modalities, such as thermal imaging and ultrasound have been also adopted in recent years due to their noninvasive nature. Finally, microscope-based assessment of the biopsy sample also is carried out to detect the cancer stage and its severity.

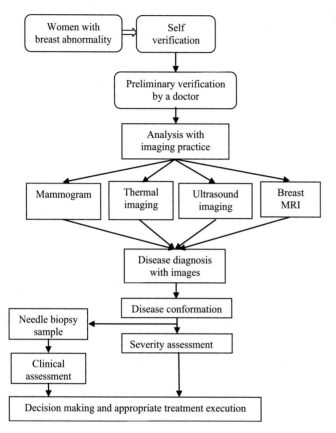

Figure 1.1 Different stages in breast cancer diagnosis.

1.2.1.1 Digital mammogram

This is a commonly used low-cost imaging process to record and inspect breast abnormalities. In this scheme, a radiograph (X-ray) is used to record the breast area using a conventional film-based or digital procedure. Recently, the existing computerized mammogram scheme has helped in achieving accurate image recording and hence it is widely used in hospitals to screen for breast abnormalities. Fig. 1.2 presents sample breast mammogram images from the mini-MIAS database [43]. This database consist of images with their class, such as fatty, fatty-glandular, and dense-glandular images, and these are useful to identify the BC as benign or malignant. It is an extensively used clinical-grade data set to assess the breast condition using a chosen image-processing scheme [44,45]. Every image in this data set is available with dimensions of $1024 \times 1024 \times 1$ pixels. Also, based on the need, a chosen image preprocessing and postprocessing

Figure 1.2 Sample mammogram images: (top) benign; (bottom) malignant; (A) right; (B) left.

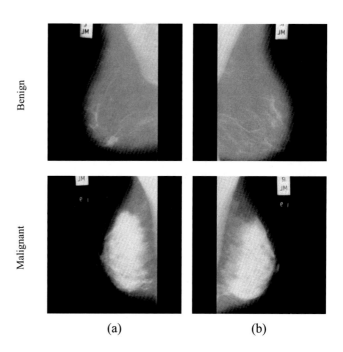

(a)   (b)

methodology can be employed to examine the BC with better accuracy. From Fig. 1.2A and B, it can be observed that the abnormal section in the breast is clearly visible and it can be easily assessed using a visual examination or an examination using the chosen computerized algorithm.

The chief advantages of mammograms include greater reliability, low cost, and easy recording. Because of this, it is one of the most commonly recommended image-assisted BC schemes by doctors. The major limitations of this scheme are as follows; it is a radiological procedure (X-ray) and it causes mild/moderate irritation to the breast area during recording. Further, it may cause mild/moderate pain in the breast when it is placed on the recording machine. Also, the recording of the mammogram may cause mild/moderate pain.

1.2.1.2 Ultrasound imaging

This is one of the safest noninvasive image-capturing techniques that is widely adopted to examine activity and disease in internal organs. For more than 2 decades, this scheme has been employed in the medical domain to record vital information from internal organs. This scheme employs a high-frequency sound signal to capture the necessary reconstructed image [46−48]. It is one of the safest approaches and does not cause tissue or organ injury. Clinical-grade breast ultrasound images can be found in Ref. [49] and the sample images collected from this data set are presented in Fig. 1.3. This figure depicts images for benign and malignant BC, and the affected area can be examined using a chosen protocol.

These images are widely adopted to record breast abnormalities in women younger than 25, with radiological allergy, or suffering from acute disease. Further, it is safe to perform on pregnant women. This scheme is identified as more accurate in detecting abnormalities in breasts with dense tissue. The common procedures followed in ultrasound imaging include: (1) the healthcare provider explains the procedure to be followed

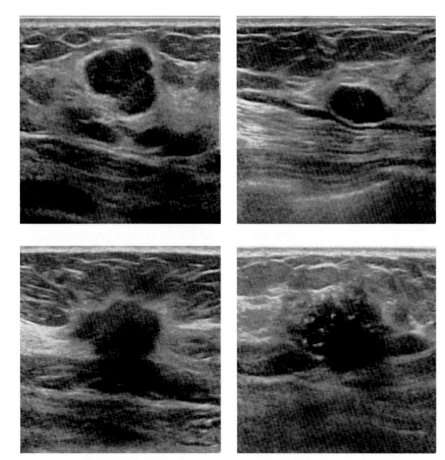

Figure 1.3 Sample breast cancer ultrasound images.

during the imaging, (2) the patient is requested to sign the consent form to perform the test, (3) the area to be examined is covered with a medical gel, and (4) a handheld ultrasound transceiver is placed close to the gel and the reflected sound wave is collected and constructed as an image using the appropriate procedure.

This technique is a straightforward and proficient method to evaluate irregularities and actions of organs with noninvasive and easy to perform imaging procedures. The main disadvantage is that the visibility of the BC is extremely poor compared with breast MRI.

1.2.1.3 Thermal imaging

The recently developed noninvasive imaging technique called breast thermal imaging (BTI) has been widely adopted to detect various breast-related abnormalities using a specially designed digital camera. This camera captures the infrared radiation (IR) emitted by the skin surface and reconstructs the IR radiation into images with varied patterns. BTI can be recorded as grayscale/RGB images and, by simply evaluating the thermal patterns, it is possible to detect abnormalities in the breast.

Fig. 1.4 depicts a grayscale BTI reproduced from Ref. [50]. After recording the image, the chosen image-processing scheme is adopted to detect the abnormality. The RGB version of the BTI is shown in Fig. 1.5. In this image, the volunteers were requested to face the camera allowing the images to be captured at different angles. After collecting these images, it was possible to detect the abnormality with a visual examination by a doctor or computer software.

Fig. 1.6 depicts the RGB and grayscale versions of images collected from the same patient. Compared to other imaging approaches, such as mammography and ultrasound, it provides clear information about the abnormal breast area including the DCIS. The mammography and ultrasound techniques are used only to record the breast mass and cannot be used to record early-stage cancer, like DCIS.

Earlier research work on thermal imaging-based breast abnormality detection can be found in Refs. [38,39,51]. From these works, it was confirmed that BTI helps to provide accurate BC detection using images recorded with grayscale/RGB images.

1.2.1.4 Breast MRI

MRI is one of the most widely adopted radiological procedures to examine disease in vital internal organs. This approach is also a

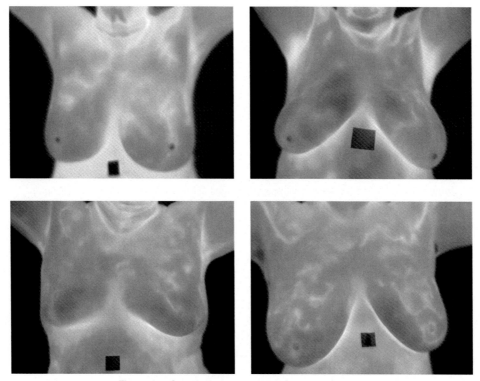

Figure 1.4 Sample breast cancer ultrasound images.

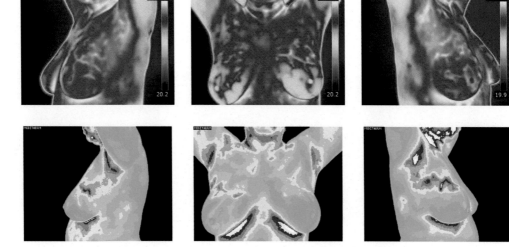

Figure 1.5 Sample breast cancer ultrasound images [50] and [51].

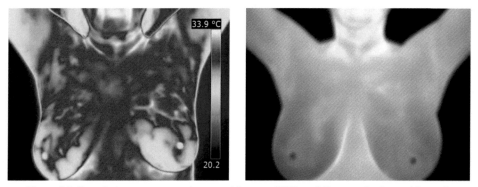

Figure 1.6 Sample breast cancer ultrasound images (RGB and Gray scale thermal image).

noninvasive technique used to record the organ of interest in three-dimensional (3D) form. The added advantages of breast MRI compared to other imaging modalities are its visibility and 3D nature [52,53]. A 3D MRI can be separated into various two-dimensional (2D) slices, which helps to obtain a clear idea of the breast abnormalities. Further, the 3D slices can be examined from various views, such as axial, coronal, and sagittal. Fig. 1.7 depicts sample 2D MRI slices of a sample breast MRI. Fig. 1.7A–C present 2D slices, and the tumor in each slice is clearly visible in all three images.

The advantages of breast MRI are as follows: (1) widely adopted radiological technique to examine the breast area, (2) provides highly visible abnormality information in 3D images, and (3) enhances disease diagnosis compared to other breast imaging modalities. The main disadvantage of MRI is that it is a radiological procedure and normally it must be executed along with a contrast agent, which may create moderate to severe side effects for the patient.

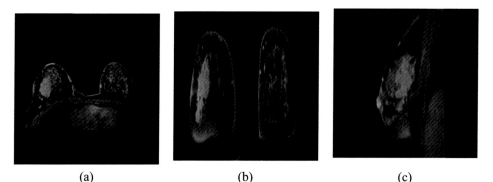

(a)   (b)   (c)

Figure 1.7 Sample 2D slices of a breast MRI: (A) axial; (B) coronal; (C) sagittal.

1.2.1.5 Breast histology

After detecting any breast abnormalities using the chosen procedure (visual inspection and image-based detection), a needle biopsy is then employed to collect tissue samples from the infected region. The collected samples are then examined using digital microscopy, which helps to categorize the infected section into benign or malignant. When benign it is the early phase of cancer and can be treated with appropriate medication, and when the abnormality is identified as malignant, surgery followed by radiotherapy is suggested. This procedure is called histology-supported diagnosis in which the microanatomy of different cells, tissues, and organs are inspected using a microscope. This method examines the connection between arrangement and functional dissimilarity among the healthy and diseased samples collected from patients [54].

Sample microscopic images are presented in Fig. 1.8 in which Fig. 1.8A presents the benign class and Fig. 1.8B shows the malignant class image.

The main advantage of a histology image is that tissue-stage analysis is possible to confirm the category and severity of the BC. The disadvantages of this technique include: sample collection from an abnormal area using a needle, preparation of a histological slide, and examination and confirmation are required by an experienced clinician.

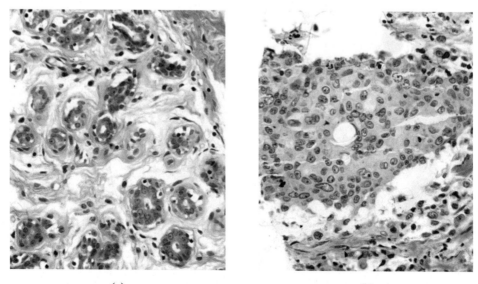

(a)   (b)

Figure 1.8 Breast histology images: (A) benign; (B) malignant.

1.2.2 Heart abnormality detection

The heart is one of the primary vital internal organs and is responsible for circulating the blood throughout the body with the help of a pumping action. For a variety of reasons, the heart abnormalities can arise, and untreated heart disease will often lead to death. In humans, heart abnormalities are generally due to the following causes:

- Coronary artery disease: This problem arises due to the deposit of cholesterol plaques on the blood valves. When the deposit level increases, the opening of the valve decreases creating a temporary/permanent valve blockage which can be cured with the appropriate medication procedures, including surgery.
- Stable angina pectoris: This is caused by a reduced coronary blood vessel which causes chest pain. Due to this problem, the exchange of oxygen from the lungs to the blood is disturbed and it may also cause notable heart disease.
- Unstable angina pectoris: This may cause moderate to severe chest pain and is an early warning for cardiac arrest. When this symptom is experienced by the patient, the patient is will need the necessary medical assistance to return the heart rhythm to its normal state.
- Myocardial infarction: This is mainly due to a problem in the coronary artery and may lead to a sudden heart attack.
- Arrhythmia: This is mainly due to a change in the normal heart beat. This problem can be identified by recording the heart's electrical activity using a chosen signal registration mechanism or recording and analyzing the heart image. The most common diagnosis involves analyzing the pattern of the ECG with a visual check and computerized algorithms.

The heart is a muscular organ and heart disease must be diagnosed in its premature phase using a suitable clinical protocol. Heart disease can be diagnosed with biosignal (ECG)- and bioimage-based methods. The use of ECG needs expertise and, in recent years, along with signal supported scheme, the image-assisted procedures have been widely adopted in hospitals for timely detection and treatment planning processes.

Fig. 1.9 depicts normal and abnormal heart activity (ECG) recorded using a single-channel signal-recording facility. Fig. 1.9A depicts the normal ECG signal and Fig. 1.9B presents an ECG from an abnormal patient. This information is recorded using surface electrodes and the information regarding this signal can be found in Ref. [55]. Earlier works related to ECG examination can be found in Refs. [24–27]. The earlier works confirmed

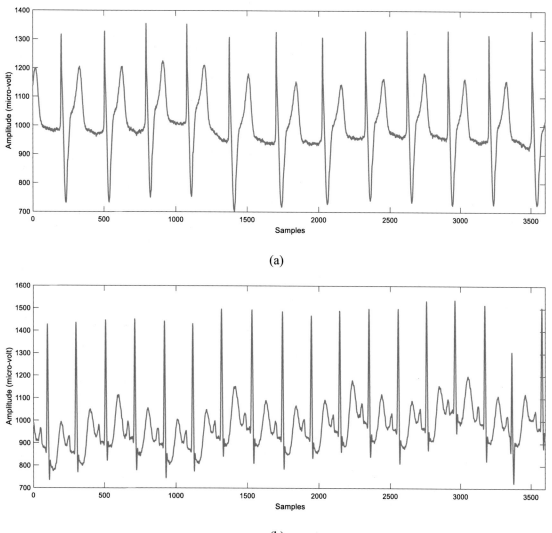

Figure 1.9 Sample ECG signals: (A) normal; (B) abnormal.

that ECG examination can be done using a specially designed signal-processing method. These works also confirmed that the information existing in the signal-based methods are much less compared to an image-based examination.

Assessment of the normal and abnormal ECG is quite complex due to the similarities in its pattern. Normally, the amplitude and frequency of this signal are computed to detect any abnormalities.

Along with the ECG-supported assessment, MRI-based heart disease detection is also widely employed in the healthcare domain. The main advantages of heart MRI are its clarity and the involvement of the contrast agent (gadolinium) during MRI recording helps to obtain better visibility of the section of the heart to be examined.

Fig. 1.10 depicts the general heart disease examination procedures followed in clinics, and the final reports attained with these procedure are then examined by an experienced doctor, who suggests the most appropriate treatment procedure(s) to treat the disease.

Fig. 1.11 depicts the benchmark heart MRI of the HVSMR 2016 database. This database has been widely examined in the literature using a considerable number of image-processing schemes,

Figure 1.10 Heart disease examination technique followed in hospitals.

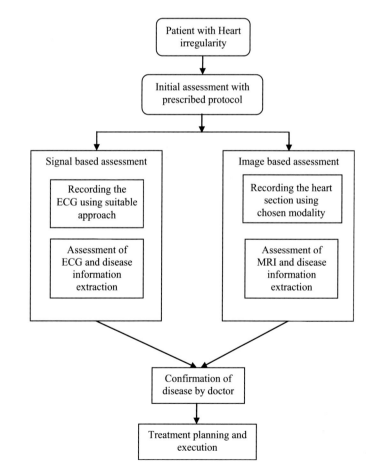

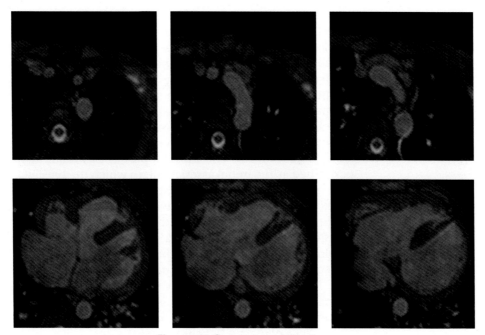

Figure 1.11 Sample heart MRI slices.

and the earlier works can be found in Ref. [56]. To find an abnormality using MRI, axial image slices are more widely adopted than other views. In this image, the highly visible section is extracted and evaluated to detect the heart abnormality.

The assessment of the heart MRI can be used to examine problems with the heart valves, blood vessels, and other heart areas with better accuracy. The earlier research works employed to extract and evaluate abnormal heart areas can be found in Refs. [57,58]. These works confirmed that MRI-supported assessment helped to localize the abnormal heart area with improved accuracy.

1.2.3 Brain abnormality detection

The brain is the major part in human physiology and is responsible for processing all signals received from other sensory body parts. After processing the signal, it will make the necessary decision by sending the response/control signal to activate or regulate the body activity. The brain abnormality will severely affect the brain's signal processing and decision-making function. Brain abnormalities are mainly due to aging, infection, injury, and other associated problems. Any abnormality in brain must be diagnosed and treated in its premature phase as delayed detection can cause severe problems. As per the literature, the most commonly found brain abnormalities include tumor, stroke, Alzheimer's, schizophrenia, multiple sclerosis, etc.

Usually, a brain abnormality will be suspected based on the symptoms experienced by the patient. After experiencing the symptoms, the patient would approach a doctor for further examination and, based on the suggestion by the doctor, the abnormality can be assessed using biosignal (EEG) and bioimage-assisted procedures. The EEG supported technique is normally performed as the initial assessment, in which the biosignal generated in the brain is collected using a scalp electrode positioned based on the prescribed pattern; complete information regarding the EEG collection can be found in Refs. [58−60].

After collecting the EEG, a suitable methodology can be employed to evaluate the EEG pattern and, after the evaluation, it is compared with the normal pattern to identify any abnormalities. The collected EEG is highly nonlinear in nature and a signal to image-supported assessment is also widely used to identify the abnormality as described in earlier research works [28,29].

Fig. 1.12 depicts a benchmark EEG signal collected from the Bonn-EEG database. This database consist of a single-channel signal of normal and epileptic (abnormal) class signals for assessment. Fig. 1.12A depicts a normal class signal and Fig. 1.12B depicts an abnormal class. Examination of the amplitude, frequency, and pattern of the EEG is necessary to detect the type and severity of the abnormality. Further, the assessment of an EEG using 1D signal is quite difficult due to its nonlinear nature. If the 1D signal is converted into a 2D image, it is possible to evaluate the abnormality with greater accuracy. The 2D images were developed using the Gramian angular summation field (GASF) technique and in this the normal/abnormal EEG signal samples are modified into 2D images, as depicted in Fig. 1.13 for normal and epileptic classes. An assessment of the GASF pattern can be found in the literature and this technique helps to obtain better disease detection compared to a 1D EEG signal.

The bioimage-based assessment of the brain is a common clinical methodology in which the patient's head area is examined using a radiological procedure. The implementation of the radiology-assisted diagnosis consists of the following steps: (1) when the patient consults the doctor with a symptom, the doctor will carry out the recommended healthcare procedures to identify the cause(s) of the symptom(s), (2) initially, an EEG-supported brain condition assessment is performed and, based

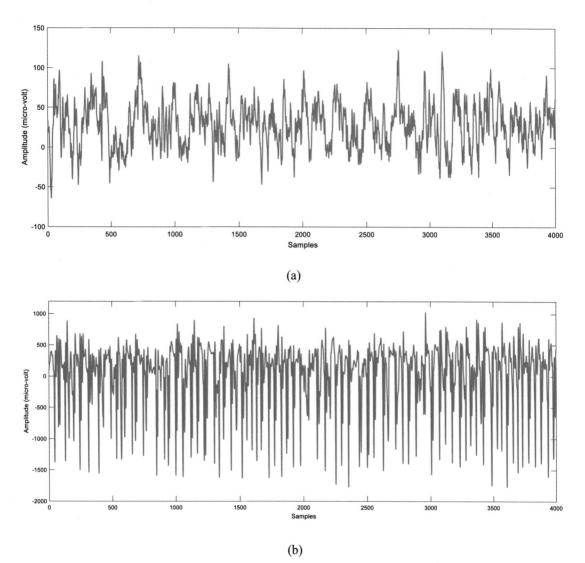

(a)

(b)

Figure 1.12 Sample single-channel EEG signal: (A) normal; (B) abnormal.

on the obtained result, the imaging procedure is recommended to get clear details of the brain condition, (3) the appropriate imaging procedure (CT/MRI) is implemented based on the suggestion of the doctor and this imaging can be implemented with/without a contrast agent, and (4) after collecting the radiological image and observation by the radiologist, the doctor once again will verify the image and finally plan the most appropriate treatment.

Normal

Epileptic

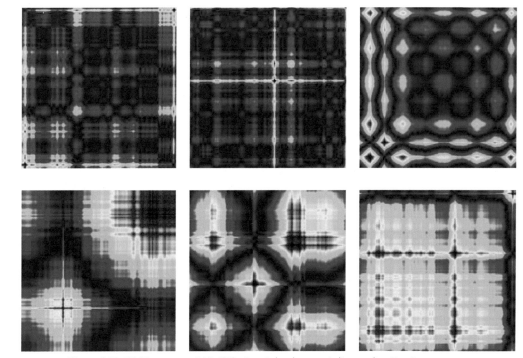

Figure 1.13 GASF pattern of 1D EEG signal; (top) normal; (bottom) epileptic.

CT and MRI are the most commonly adopted imaging schemes to record the brain for further examination. In CT, the bone section and the section enhanced by the contrast agent will have better visibility compared to the gray matter, white matter, and cerebrospinal fluid. A sample brain CT is depicted in Fig. 1.14 with axial and coronal views. This image confirms that the disease information seen is poorer using CT than MRI.

Fig. 1.15 presents brain MRI slices of ischemic stroke lesion segmentation (ISLES2015) challenge data set images widely used in the literature [61−63]. All the images are available in 3D form and the added advantage of MRI compared to other imaging schemes is that it can be recorded with a range of modalities, such as T1-weighted (T1), T1-contrast enhanced (T1C), T2-weighted (T2), flair, and diffusion-weight (DW).

The MRI in ISLES2015 is recorded with modalities, such as DW, flair, and T1, as shown in Fig. 1.15. Every image has its advantages and disadvantages. In DW and flair modalities, the stroke section is highly visible compared to the normal brain section, and in the T1 case, the visibility of the infected section is very

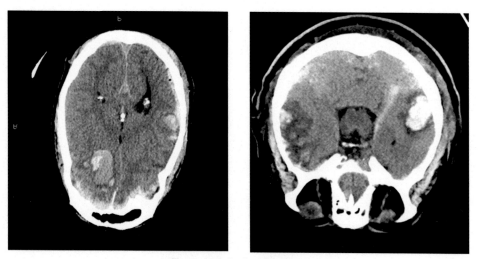

Figure 1.14 Brain CT images.

poor. In most MRIs, the brain section is associated with a skull fragment and the clinical level assessment of the MRI can be performed with/without the skull. Normally, the skull in a brain MRI is removed using a threshold filter, as discussed in earlier works [30,32].

The 3D MRI can be separated into different 2D slices, as presented in Fig. 1.15. Fig. 1.15A−C present the axial, coronal, and sagittal views, respectively, and during the examination, any one or all of the image views can be considered. However, normally the clinical-level assessment is performed by a doctor using the axial view, and when planning the surgery (keyhole/open skull), the doctor will examine all three views carefully.

Because of its advantages, MRI is one of the most widely adopted imaging procedures used to examine the condition of the brain with better accuracy. The earlier works which employed the MRI method to assess various brain abnormalities can be found in Ref. [64]. The earlier works implemented the following procedures using brain MRI slices: abnormality localization, segmentation, machine learning (ML)-based classification, and deep learning (DL)-supported detection [4,30,32−34]. Most of these works employed 2D slices because of their simplicity and very few works examined brain abnormalities using 3D whole MRI images, which needs a more complex methodology compared to the traditional 2D slice examination.

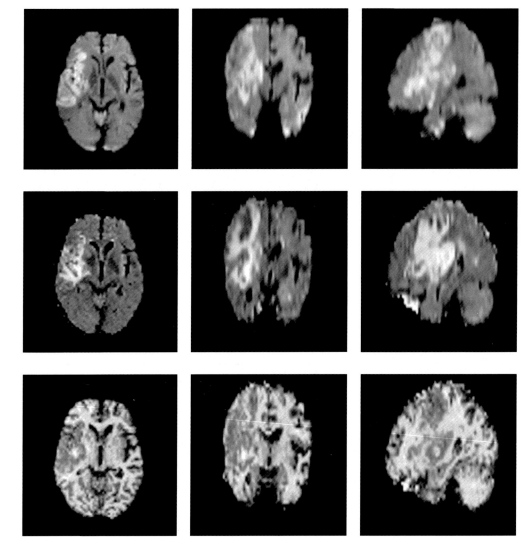

Figure 1.15 Brain MRI slices of varied modalities: (top) MR; (center) flair; (bottom) T1; (A) axial; (B) coronal; (C) sagittal.

The methodology adopted in clinical level brain abnormalities is presented in Fig. 1.16. The patient approaches a doctor with an evident symptom for further assessment. The doctor will initially suggest a biosignal-based assessment due to its safety compared to radiological imaging schemes. When the EEG examination

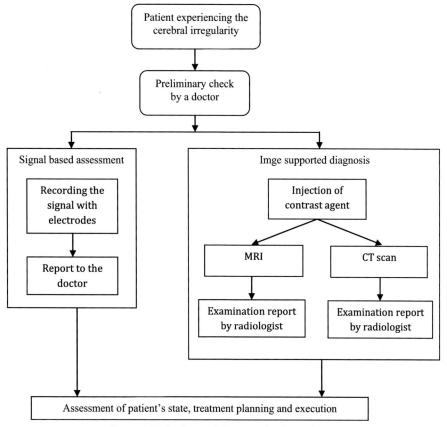

Figure 1.16 Brain condition examination scheme.

provides a suspicious signal pattern, then they will suggest a radiology-supported image examination with/without a contrast agent. The suggestion to use a contrast agent along with its dose level must be considered based on the patient's health condition, as the contrast agent can cause mild to moderate side effects. From this figure, it can be noted that the choice of imaging scheme (CT/MRI) depends mainly on the brain abnormality to be monitored and, in most disease cases, MRI is more commonly recommended than CT due to its advantages, such as clarity of the affected area, multiple modalities, and reliability. The combined results of the personal examination, EEG assessment report, radiologist findings, and the doctor's own findings from the CT/MRI influence the choice of treatment.

1.3 Discussion of the need for an imaging scheme

Biomedical images are widely preferred to examine human disease compared with biosignals due to their simplicity and clarity. The choice of imaging modality depends on the disease to be examined, the organ which is affected, the patient's health condition, and the expertise of the doctor. This chapter examines the cases of diseases in vital organs, including the breasts, heart, and brain.

Earlier work on breast abnormality assessment (BC) confirmed that disease in the breasts can be easily assessed with imaging approaches, such as mammography, thermal imaging, and ultrasound. Although these images provide vital information regarding the disease, doctors require a clear picture of the type of disease, its location, and severity. To obtain complete information about a breast abnormality, breast MRIs are widely adopted by researchers, and the earlier research works with MRI-supported abnormality assessment can be found in Ref. [65]. MRI provides a 3D picture, which gives sufficient information regarding the disease, to enable the treatment to be planned without confirming it with a needle biopsy test.

Previously, the assessment heart abnormalities was widely performed using ECG, in which the electrical activity of the heart can be monitored by analyzing the ECG pattern and the change in pattern can be used to examine the specific abnormality. After identifying the abnormality in the heart, its condition must be checked using an open heart surgery (OHS) procedure, which is an invasive approach. After surgery, the cause of the heart abnormality can be identified and treated. Recently, heart MRI has been considered to obtain the complete information about the heart in a reconstructed 3D image. Without surgery, the doctor is able to perform an analysis on the MRI to identify the abnormality. After detecting the abnormality, the heart specialist will decide the treatment to be executed to treat the disease, and this can include: a recommended medication, minor surgery, or OHS. Further, recovery of the heart after treatment can be predicted using MRI images.

The clinical-level assessment of brain conditions can be performed using signal (EEG)- and image (CT/MRI)-based procedures. EEG assessment is moderately difficult due to its nonlinearity and as it requires clinician expertise. Hence, most brain abnormalities are examined using an image-supported scheme and, particularly, suspicious brain areas are recorded and examined using MRI. MRI has the added advantage compared to CT that it enables higher visibility, varied modalities, etc. The information available in an MRI can be easily identified by a radiologist and/or doctor compared to other brain examination approaches. Along with MRI, another variant called the magnetic resonance angiogram (MRA) is also widely adopted to examine specific areas of the brain (blood vessels or tissue). The earlier research works on brain MRI have confirmed its advantages over other examination procedures. Detailed information regarding MRI-assisted brain assessment can be found in Refs. [66–68].

1.4 Summary

In the current era, the incidence rate diseases has increased due to various reasons. In order to identify a disease in its premature phase, a number of diagnostic procedures are proposed and implemented to reduce the mortality rate. Further, modern hospitals are equipped with remote patient monitoring facilities, which help the patient to follow the appropriate guidance to enhance their recovery rate. This chapter has presented an overview of the imaging procedures considered to examine abnormalities in organs, such as the breasts, heart, and brain. For each of these organs, the possible disease detection procedure, its advantages and disadvantages are discussed with appropriate examples, and the need for MRI-assisted diagnosis is highlighted.

Compared to the other imaging schemes, MRI has considerable merits, such as multiple modalities (T1, T1C, T2, flair, and DW), multiple views (axial, coronal, and sagittal), and flexibility to examine the abnormality by choosing 3D and 2D images, etc. Compared to other disease diagnostic procedures, the MRI-supported procedures have been more widely adopted in recent times due to their reliability. MRI can be recorded using conventional methods (transparent film) and a modern approach (digital images). Further, this radiological method helps to obtain the initial report by the radiologist and a final report by the doctor. The combination of these two reports will help to get a better understanding of the nature of the disease, its location, and severity. Finally, the doctor will plan the possible treatment for the patient.

References

[1] Rajinikanth V, Priya E, Lin H, Lin F. Hybrid image processing methods for medical image examination. CRC Press; 2021.
[2] Das H, Dey N, Balas VE, editors. Real-time data analytics for large scale sensor data. Academic Press; 2019.
[3] Priya E, Rajinikanth V. Signal and image processing techniques for the development of intelligent healthcare systems. Springer; 2020.

[4] Fernandes SL, Tanik UJ, Rajinikanth V, Karthik KA. A reliable framework for accurate brain image examination and treatment planning based on early diagnosis support for clinicians. Neural Computing and Applications 2020; 32(20):15897−908.

[5] Acharya UR, Fernandes SL, WeiKoh JE, Ciaccio EJ, Fabell MKM, Tanik UJ, et al. Automated detection of Alzheimer's disease using brain MRI images−a study with various feature extraction techniques. Journal of Medical Systems 2019;43(9):1−14.

[6] Qureshi F, Krishnan S. Wearable hardware design for the internet of medical things (IoMT). Sensors 2018;18(11):3812.

[7] Vishnu S, Ramson SJ, Jegan R. Internet of medical things (IoMT)-An overview. In: 2020 5th international conference on devices, circuits and systems (ICDCS). IEEE; 2020 March. p. 101−4.

[8] Kumari A, Tanwar S, Tyagi S, Kumar N. Fog computing for healthcare 4.0 environment: opportunities and challenges. Computers & Electrical Engineering 2018;72:1−13.

[9] Dubey H, Monteiro A, Constant N, Abtahi M, Borthakur D, Mahler L, et al. Fog computing in medical internet-of-things: architecture, implementation, and applications. In: Handbook of large-scale distributed computing in smart healthcare. Cham: Springer; 2017. p. 281−321.

[10] Dey N, Das H, Naik B, Behera HS, editors. Big data analytics for intelligent healthcare management. Academic Press; 2019.

[11] Dey N, Tamane S, editors. Big data analytics for smart and connected cities. IGI Global; 2018. p. 348.

[12] Harerimana G, Jang B, Kim JW, Park HK. Health big data analytics: a technology survey. IEEE Access 2018;6:65661−78.

[13] Chaitanya NK, Radhakrishnan A, Reddy GR, Manikandan MS. A simple and robust QRS detection algorithm for wireless medical body area network. In: 2011 international conference on emerging trends in networks and computer communications (ETNCC). IEEE; 2011, April. p. 153−8.

[14] Fatima M, Kiani AK, Baig A. Medical body area network, architectural design and challenges: a survey. In: Wireless sensor networks for developing countries. Berlin, Heidelberg: Springer; 2013, April. p. 60−72.

[15] Jin Y. Low-cost and active control of radiation of wearable medical health device for wireless body area network. Journal of Medical Systems 2019; 43(5):1−11.

[16] Aida A, Svensson T, Svensson AK, Urushiyama H, Okushin K, Oguri G, et al. Using mHealth to provide mobile app users with visualization of health checkup data and educational videos on lifestyle-related diseases: methodological framework for content development. JMIR mHealth and uHealth 2020;8(10):e20982.

[17] Rajinikanth V, Kadry S. Development of a framework for preserving the disease-evidence-information to support efficient disease diagnosis. International Journal of Data Warehousing and Mining (IJDWM) 2021;17(2): 63−84.

[18] Morse SS. Factors in the emergence of infectious diseases. Plagues Politics 2001:8−26.

[19] Jones KE, Patel NG, Levy MA, Storeygard A, Balk D, Gittleman JL, et al. Global trends in emerging infectious diseases. Nature 2008;451(7181):990−3.

[20] Fauci AS. Infectious diseases: considerations for the 21st century. Clinical Infectious Diseases 2001;32(5):675−85.

[21] Dagger TS, Sweeney JC, Johnson LW. A hierarchical model of health service quality: scale development and investigation of an integrated model. Journal of Service Research 2007;10(2):123–42.

[22] Boyd MR, Powell BJ, Endicott D, Lewis CC. A method for tracking implementation strategies: an exemplar implementing measurement-based care in community behavioral health clinics. Behavior Therapy 2018;49(4): 525–37.

[23] Chib A, van Velthoven MH, Car J. mHealth adoption in low-resource environments: a review of the use of mobile healthcare in developing countries. Journal of Health Communication 2015;20(1):4–34.

[24] Martis RJ, Acharya UR, Adeli H. Current methods in electrocardiogram characterization. Computers in Biology and Medicine 2014;48:133–49.

[25] Acharya UR, Chua ECP, Faust O, Lim TC, Lim LFB. Automated detection of sleep apnea from electrocardiogram signals using nonlinear parameters. Physiological Measurement 2011;32(3):287.

[26] Dey N, Ashour AS, Shi F, Fong SJ, Sherratt RS. Developing residential wireless sensor networks for ECG healthcare monitoring. IEEE Transactions on Consumer Electronics 2017;63(4):442–9.

[27] Dey N, Samanta S, Yang XS, Das A, Chaudhuri SS. Optimisation of scaling factors in electrocardiogram signal watermarking using cuckoo search. International Journal of Bio-Inspired Computation 2013;5(5):315–26.

[28] Jahmunah V, Oh SL, Rajinikanth V, Ciaccio EJ, Cheong KH, Arunkumar N, Acharya UR. Automated detection of schizophrenia using nonlinear signal processing methods. Artificial Intelligence in Medicine 2019;100:101698.

[29] Thanaraj KP, Parvathavarthini B, Tanik UJ, Rajinikanth V, Kadry S, Kamalanand K. Implementation of deep neural networks to classify EEG signals using gramian angular summation field for epilepsy diagnosis. arXiv 2020;2003:04534. preprint arXiv.

[30] Rajinikanth V, Satapathy SC, Fernandes SL, Nachiappan S. Entropy based segmentation of tumor from brain MR images—a study with teaching learning based optimization. Pattern Recognition Letters 2017;94:87–95.

[31] Raja NSM, Fernandes SL, Dey N, Satapathy SC, Rajinikanth V. Contrast enhanced medical MRI evaluation using Tsallis entropy and region growing segmentation. Journal of Ambient Intelligence and Humanized Computing 2018:1–12.

[32] Rajinikanth V, Satapathy SC. Segmentation of ischemic stroke lesion in brain MRI based on social group optimization and Fuzzy-Tsallis entropy. Arabian Journal for Science and Engineering 2018;43(8):4365–78.

[33] Rajinikanth V, Fernandes SL, Bhushan B, Sunder NR. Segmentation and analysis of brain tumor using Tsallis entropy and regularised level set. In: Proceedings of 2nd international conference on micro-electronics, electromagnetics and telecommunications. Singapore: Springer; 2018. p. 313–21.

[34] Rajinikanth V, Dey N, Satapathy SC, Ashour AS. An approach to examine magnetic resonance angiography based on Tsallis entropy and deformable snake model. Future Generation Computer Systems 2018;85:160–72.

[35] Dey N, Zhang YD, Rajinikanth V, Pugalenthi R, Raja NSM. Customized VGG19 architecture for pneumonia detection in chest X-rays. Pattern Recognition Letters 2021;143:67–74.

[36] Holland R, Peterse JL, Millis RR, Eusebi V, Faverly D, van de Vijver MA, et al. Ductal carcinoma in situ: a proposal for a new classification. In: Seminars in diagnostic pathology, vol. 11; August 1994. p. 167–80. 3.

[37] Groen EJ, Elshof LE, Visser LL, Emiel JT, Winter-Warnars HA, Lips EH, et al. Finding the balance between over-and under-treatment of ductal carcinoma in situ (DCIS). The Breast 2017;31:274–83.

[38] Raja N, Rajinikanth V, Fernandes SL, Satapathy SC. Segmentation of breast thermal images using Kapur's entropy and hidden Markov random field. Journal of Medical Imaging and Health Informatics 2017;7(8):1825–9.

[39] Fernandes SL, Rajinikanth V, Kadry S. A hybrid framework to evaluate breast abnormality using infrared thermal images. IEEE Consumer Electronics Magazine 2019;8(5):31–6.

[40] Rajinikanth V, Raja NSM, Satapathy SC, Dey N, Devadhas GG. Thermogram assisted detection and analysis of ductal carcinoma in situ (DCIS). In: 2017 international conference on intelligent computing, instrumentation and control technologies (ICICICT). IEEE; 2017, July. p. 1641–6.

[41] Nair MV, Gnanaprakasam CN, Rakshana R, Keerthana N, Rajinikanth V. Investigation of breast melanoma using hybrid image-processing-tool. In: 2018 international conference on recent trends in advance computing (ICRTAC). IEEE; September 2018. p. 174–9.

[42] https://www.who.int/news-room/fact-sheets/detail/breast-cancer.

[43] http://peipa.essex.ac.uk/info/mias.html.

[44] Zemmal N, Azizi N, Dey N, Sellami M. Adaptive semi supervised support vector machine semi supervised learning with features cooperation for breast cancer classification. Journal of Medical Imaging and Health Informatics 2016;6(1):53–62.

[45] Virmani J, Dey N, Kumar V. PCA-PNN and PCA-SVM based CAD systems for breast density classification. In: Applications of intelligent optimization in biology and medicine. Cham: Springer; 2016. p. 159–80.

[46] Sehgal CM, Weinstein SP, Arger PH, Conant EF. A review of breast ultrasound. Journal of Mammary Gland Biology and Neoplasia 2006;11(2): 113–23.

[47] Weinstein SP, Conant EF, Sehgal C. Technical advances in breast ultrasound imaging. In: Seminars in ultrasound, CT and MRI, vol. 27. WB Saunders; August 2006. p. 273–83. 4.

[48] Thanaraj RIR, Anand B, Rahul JA, Rajinikanth V. Appraisal of breast ultrasound image using Shannon's thresholding and level-set segmentation. In: Progress in computing, analytics and networking. Singapore: Springer; 2020. p. 621–30.

[49] http://www.onlinemedicalimages.com/index.php/en/.

[50] http://visual.ic.uff.br/dmi/#:~:text=Database%20of%20mastologic% 20images,obtained%20by%20our%20research%20group.

[51] Dey N, Rajinikanth V, Hassanien AE. An examination system to classify the breast thermal images into early/acute DCIS class. In: Proceedings of international conference on data science and applications. Singapore: Springer; 2021. p. 209–20.

[52] Clark K, Vendt B, Smith K, Freymann J, Kirby J, Koppel P, et al. The Cancer Imaging Archive (TCIA): maintaining and operating a public information repository. Journal of digital imaging 2013;26(6):1045–57.

[53] Meyer CR, Chenevert TL, Galban CJ, Johnson TD, Hamstra DA, Rehemtulla A, et al. Data from RIDER_Breast_MRI. The Cancer Imaging Archive 2015. https://doi.org/10.7937/K9/TCIA.2015.H1SXNUXL.

[54] Aksac A, Demetrick DJ, Ozyer T, et al. BreCaHAD: a dataset for breast cancer histopathological annotation and diagnosis. BMC Research Notes 2019;12:82.

[55] Plawiak P. ECG signals (1000 fragments). Mendeley Data, v3; 2017.

[56] Yu L, Cheng JZ, Dou Q, Yang X, Chen H, Qin J, et al. Automatic 3D cardiovascular MR segmentation with densely-connected volumetric convNets. In: International conference on medical image computing and computer-assisted intervention. Cham: Springer; September 2017. p. 287–95.

[57] Mukhopadhyay A. Total variation random forest: fully automatic mri segmentation in congenital heart diseases. In: Reconstruction, segmentation, and analysis of medical images. Cham: Springer; 2016. p. 165–71.

[58] Lin H, Rajinikanth V. Development of softcomputing tool to evaluate heart MRI slices. International Journal of Computer Theory and Engineering 2019;11(5):80–3.

[59] Le Van Quyen M, Martinerie J, Navarro V, Boon P, D'Havé M, Adam C, et al. Anticipation of epileptic seizures from standard EEG recordings. The Lancet 2001;357(9251):183–8.

[60] Boon P, Michielsen G, Goossens L, Drieghe C, D'Have M, Buyle M, et al. Interictal and ictal video-EEG monitoring. Acta Neurologica Belgica 1999; 99(4):247–55.

[61] Maier O, Menze BH, von der Gablentz J, Häni L, Heinrich MP, Liebrand M, et al. ISLES 2015-A public evaluation benchmark for ischemic stroke lesion segmentation from multispectral MRI. Medical Image Analysis 2017;35: 250–69.

[62] Carass A, Roy S, Gherman A, Reinhold JC, Jesson A, Arbel T, et al. Evaluating white matter lesion segmentations with refined Sørensen-Dice analysis. Scientific Reports 2020;10(1):1–19.

[63] Winzeck S, Hakim A, McKinley R, Pinto JA, Alves V, Silva C, et al. ISLES 2016 and 2017-benchmarking ischemic stroke lesion outcome prediction based on multispectral MRI. Frontiers in Neurology 2018;9:679.

[64] Weber S, Habel U, Amunts K, Schneider F. Structural brain abnormalities in psychopaths—a review. Behavioral Sciences and the Law 2008;26(1):7–28.

[65] DeMartini W, Lehman C, Partridge S. Breast MRI for cancer detection and characterization: a review of evidence-based clinical applications. Academic Radiology 2008;15(4):408–16.

[66] Balafar MA, Ramli AR, Saripan MI, Mashohor S. Review of brain MRI image segmentation methods. Artificial Intelligence Review 2010;33(3):261–74.

[67] Wadhwa A, Bhardwaj A, Verma VS. A review on brain tumor segmentation of MRI images. Magnetic Resonance Imaging 2019;61:247–59.

[68] Lladó X, Oliver A, Cabezas M, Freixenet J, Vilanova JC, Quiles A, et al. Segmentation of multiple sclerosis lesions in brain MRI: a review of automated approaches. Information Sciences 2012;186(1):164–85.

Further reading

[1] http://www.thermography.co.in/.

[2] Chowdhury L, Chowdhury BR, Rajinikanth V, Dey N. A framework to evaluate and classify the clinical-level EEG signals with epilepsy. In: Proceedings of international conference on data science and applications. Singapore: Springer; 2021. p. 111–21.

2

Magnetic resonance imaging: recording and reconstruction

2.1 Introduction

In recent years, medical images have played an increasingly vital role in detecting infectious and acute diseases in humans. The recent advancements in imaging techniques have helped to make diagnosis very simple and most of the imaging approaches provide a digital image of the section/part under diagnosis. Digital images collected from patients can be easily stored, processed, and examined in a specified location (computer memory). In the literature, a considerable number of noninvasive imaging methods have been proposed and adopted in the healthcare domain for the examination of human diseases [1−3]. Magnetic resonance imaging (MRI) is one of the most widely adopted radiology-supported imaging methods, particularly for use in detecting abnormalities in internal body parts. The multiple modality and capability of producing 3D images make MRI the perfect choice compared to other techniques. The recorded MRIs can be examined in 3D or 2D formats and the visibility of the disease section (lesion/infection) is greater in MRI compared with other imaging schemes such as computed tomography (CT), thermal imaging, and ultrasound [4−8].

MRI is widely recommended by doctors to examine cancer and its severity in selected organs. When a patient approaches a doctor with a disease symptom, the doctor will initially perform a detailed health examination on the patient to identify the cause and severity of disease. When the disease is located at a sensitive body organ, the doctor then will recommend examination of the organ using MRI. After obtaining the MRI slice, the doctor examines the MRI slides and identifies the cause and nature of the illness. MRI is generally recommended to examine disease in vital parts, such as the breasts, heart, and brain. Particularly, MRI is recommended when the patient is suffering due to the impact of cancer.

Magnetic Resonance Imaging: Recording, Reconstruction and Assessment. https://doi.org/10.1016/B978-0-12-823401-3.00003-1

Table 2.1 presents information about MRI-supported abnormality diagnostic schemes.

This chapter presents information, such as the clinical level recording of MRI, the role of the contrast agent in MRI, recording

Table 2.1 Various disease detection methods using MRI slices.

Reference	Methodology employed
Pace et al. [9]	This work proposed an algorithm to mine the heart chambers and epicardial surfaces in pediatric cardiac MRI of congenital heart disease
Zhuang and Shen [10]	A method based on a multiscale patch was developed to segment the whole heart using MRI
Mukhopadhyay [11]	Implementation of an automated supervised segmentation scheme for mining the great vessel and blood pool from the heart using MRI is discussed
Akhbardeh and Jacobs [12]	A detailed assessment of various breast MRI segmentation procedures is discussed
Kang et al. [13]	Segmentation of breast MRI using K-means clustering-based preprocessing is discussed
Fashandi et al. [14]	Segmentation of the breast MRI with the UNet scheme is demonstrated using 2D MRI slices
Gubern-Merida et al. [15]	This work implemented an automatic image examination scheme to detect the breast density in breast MRIs
Adoui et al. [16]	Segmentation of tumors in breast MRI is implemented using CNN-based encoder—decoder assembly and this work helped to mine the tumor automatically
Liu et al. [17]	Segmentation of the stroke lesion in MRI slice is presented using the deep CNN-supported segmentation scheme
Mitra et al. [18]	Localization and mining of the stroke lesion in MRI is presented using a random forest technique
Clèrigues et al. [19]	Automated segmentation of acute and subacute stroke lesions in multimodal MRI is presented and discussed using clinical-grade MRI slices
Subudhi et al. [20]	A joint segmentation and classification of stroke lesions in MRI is presented
Subudhi et al. [21]	This work presents an expectation-maximization algorithm and random forest classification-based automated detection of stroke in brain MRIs
Maier et al. [22]	A detailed assessment of classifiers employed to segment strokes in MRI slices is discussed with experimental results
Akkus et al. [23]	A detailed assessment of deep-learning-supported brain tumor examination using MRI is discussed
Gordillo et al. [24]	This work presents a detailed survey of methods employed to segment tumors in brain MRI
Pereira et al. [25]	CNN-supported automatic segmentation of tumors in brain MRI is presented
Wadhwa et al. [26]	A detailed review of automated and semiautomated brain tumor mining from MRI slices is discussed
Pereira et al. [27]	Implementation of a semantic segmentation scheme to extract the abnormal section in brain MRI is discussed using an adaptive feature recombination technique

protocol, and methods adopted in evaluating MRI. When the necessary image is attained from the patient, then it will be initially evaluated by the radiologist to compute the disease score, and then the images along with the radiologist's report are sent to the doctor for further assessment. In modern clinics, the MRI is initially examined using a chosen computer algorithm and the report generated by the computer algorithm is sent to the doctor along with the radiology report. The doctor will perform a visual check on the selected MRI slices and based on the outcome and disease score in the report, the doctor will plan for an appropriate treatment to reduce the disease impact. When the appropriate treatment is implemented, the patient is further screened using MRI based on the doctor's advice to confirm the recovery.

2.2 MRI recording protocol

When a patient is suspected to have disease in an internal body part, the doctor will recommend a bioimage screening procedure to confirm the disease, its location, and severity.

2.2.1 Preparing the patient

Examination of the patient with an MRI needs the following preparatory procedures:
- When the doctor suggests MRI screening for a patient for further examination, the patient must visit the recommended radiology center along with the recommendation letter provided by the doctor.
- The radiologist will interview the patient about the necessary information regarding their health, allergies, the earlier MRI examination implemented, etc. The radiologist/technician also requests the patient to remove any external metal ornaments. Further, a through enquiry regarding having a pacemaker, aneurysm clips, metal rods/plates in broken bones, and other surgery-related information is conducted with the patient.
- During the MRI examination, informed consent must be obtained from each patient after explaining the necessary processes to be implemented.
- The MRI examination begin with sedations under the supervision of a trained healthcare supervisor.
- After the necessary precautionary measures, the patient is taken to a controlled environment, where the MRI is recorded using a specialized imaging machine. This procedure is repeated until the necessary information is recorded using the MRI.

Detailed information regarding patient preparation for an MRI examination can be found in Refs. [28,29].

2.2.2 Contrast agent

The examination of the disease in a chosen area can be recorded with the help of contrast agents (CAs). CA are pharmaceuticals that enhance the disease information in the diagnostic bioimage. Normally, the CA helps to enhance the sensitivity and specificity of MRI by changing the essential information in tissues, which manipulates the contrast of recorded image. Planned localization of CA can help to modify the tissue information and gives the necessary improvement in the images. A non-iodine CA may be injected into the patient's blood through a vein to give enhanced images. Before injecting the contrast agent (gadolinium), it is necessary to disclose relevant information, such as kidney disease, allergic reaction to a specific drug, and other health issues. The CA and its dose level will be planned by the radiologist based on the condition of the patient.

During the examination, the contrast of recorded MRI depends on the difference in intensity between the region of interest and its background. The well-illustrated dissimilarity in intensity among diverse tissue types helps to obtain a superior enhancement of the affected section.

Gadolinium-based CA is paramagnetic, in which the atoms act like ferromagnetic and superparamagnetic substances, and includes a positive magnetic weakness. The strength of the paramagnetic material is considerably lesser than in other substances with positive susceptibility. Paramagnetic particles contain autonomous magnetically diffused moments. The provoked magnetization proceeds to zero when the functional magnetic field is turned off.

Augmented in-vivo can be obtained by an improvement in tissue's signal strength, reduction in longitudinal-relaxation-time (T1) and reduction in transverse-relaxation-time (T2). Paramagnetic particles apply their control on the MR signal based on this method, improving the effectiveness of T1 and T2 relaxation. Complete information regarding the role of the contrast agent can be found in Refs. [30−32].

2.2.3 Radiology-assisted recording

The MRI recording of a patient involves: (1) preparing the patient for a scheduled examination, (2) inserting the contrast agent (gadolinium-based) into the blood veins of the patient, and (3) executing the radiology-assisted image recording and reconstruction process.

The radiology-assisted image recording must be performed in a controlled environment under the supervision of a skilled radiologist and technician. The patient is placed on the bed of the scanning device, which helps to capture the image of a particular body area. When the patient receives a controlled magnetic field, the hydrogen atoms in the body fluid line-up with respect to the magnetic field. At this stage, when oscillatory radiofrequency pulses are sent toward the atoms, it will flip to the other plan and once again return to its original state. The returning time is called the relaxation time and this time varies based on the type and nature of the tissue. This time also can be used to differentiate normal and disease-affected tissues. Based on the relaxation time, it is commonly classified into longitudinal relaxation time (T1) and transverse relaxation time (T2) and these two conditions are widely considered to obtain the images of most body organs.

The procedure followed during the MRI recording is as follows:

- The MRI scanning machine is a customized tunnel-like setup with a moving table. When the patient lies on the paddletable, it is then moved to the appropriate position to make a complete scan of the patient's body area which is to be examined.
- The patient lies on the table until the procedure is completed. Any physical movement by the patient can cause a faulty/distorted image.
- The whole scanning time may vary from 20 to 90 min according to the body area to be examined, modalities adopted, required clarity of the image, etc.
- MRI is a noninvasive technique and does not cause any pain during the imaging procedure. The procedure helps to obtain the complete information about the body organ being examined. Normally, the obtained image is in a 3D form and it can be evaluated in 3D or 2D form. After completion of the MRI, if the patient feels any abnormality, other symptom, or pain, the doctor will thoroughly check and treat the patient with the necessary medication. This kind of mild abnormality can be experienced by patients with an allergenic reaction due to gadolinium.

2.2.4 Reconstruction of MRI

The normal/abnormal areas of the body organs are recorded using MRI of the chosen modality and, based on the modality, there is a slight/considerable change in the MRI contrast between the normal and abnormal areas. Normally, the basic modalities, such as T1, T1C, T2, flair, and DW are widely adopted to record the images. The outcome of MRI scanning is provided as a soft-copy (digital images of 3D or 2D slices) and the traditional film-based copies (2D slices only). These images are further examined by a doctor or a computer algorithm to verify and validate the abnormalities [33,34]. The various MRI modalities considered to record the images are presented in this subsection.

T1 modality: When an imaging device is set to capture the T1-weighted image, the tissues with a low T1 value produce a brighter section image because of the elevated magnetization that is recorded and reconstructed. It produces T1-associated contrast by reducing the contributions of T2, which is achieved using a short reception time. The suspicious (pathological section) region will appear dark in the T1-weighted picture, as this state usually increases the water content in the disease-affected area and this region causes a signal loss when captured with the T1 modality.

The visibility of various colors with respect to the organ parts is described in Table 2.2;

Example T1-weighted modality images are presented in Fig. 2.1.

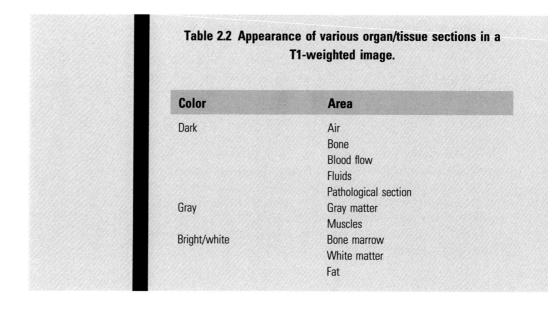

Table 2.2 Appearance of various organ/tissue sections in a T1-weighted image.

Color	Area
Dark	Air
	Bone
	Blood flow
	Fluids
	Pathological section
Gray	Gray matter
	Muscles
Bright/white	Bone marrow
	White matter
	Fat

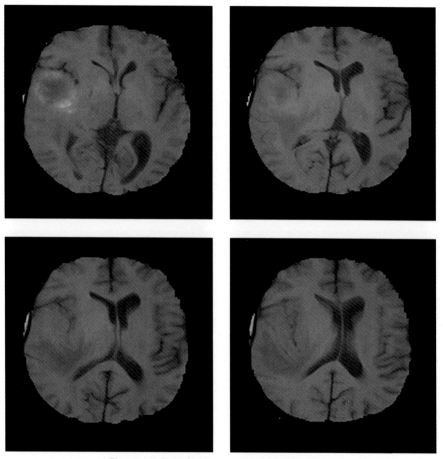

Figure 2.1 Example T1-weighted 2D MRI slices.

Let us consider the 2D brain MRI slice depicted in Fig. 2.1 of BRATS2015 [35,36] for the discussion. This image shows a lower contrast image (difference between light and dark pixels) and in which the dark, gray, and bright pixels are formed based on the amplitudes of the magnetic resonance signal. The brighter section depicts the larger amplitude, and the dark and gray show the mild and mid-amplitude levels.

In a T1-weighted scheme, the reception time (TR) indicates the quality of the recorded image. The TR is the duration of relaxation periods between two excitation pulses. Based on the TR, the T1 is grouped into low T1 (short TR) and high T1 (long TR). Other related information about the T1 and its duration can be found in Refs. [32,37].

Fig. 2.2 depicts sample test images of T1-contrast-enhanced (T1C) 2D MRI slices, in which the pathological section has more bright pixels than the T1-weighted image. The T1C is normally used to detect a tumor enhanced with the considered contrast agent.

T2 modality: When the MR signal is set to capture a T2-weighted image, the tissues with a high T2 value produce a brighter section image because of the elevated magnetization that is recorded and reconstructed. It creates T2-connected contrast by reducing the contributions of T1, which is achieved using a short reception time. The suspicious (pathological section) area will appear bright in the T2-weighted image, as

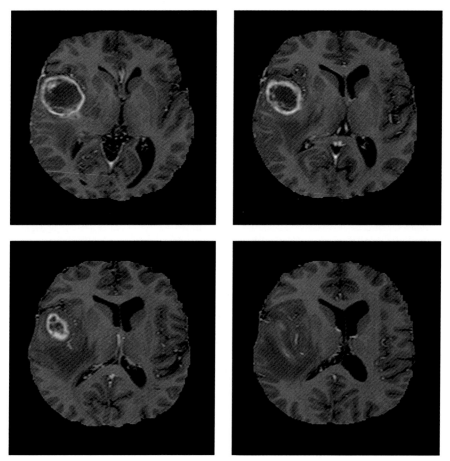

Figure 2.2 Example T1C-weighted 2D MRI slices.

Table 2.3 Appearance of various organ/tissue sections in a T2-weighted image.

Color	Area
Dark	Air
	Bone
	Blood flow
	White matter
Gray	Gray matter
	Muscles
Bright/white	Bone marrow
	Fluids
	Fat
	Pathological section

this state usually increases the water content in the disease-affected area and this region causes a signal increase when captured with the T2 modality.

Table 2.3 presents the various colors of the image pixels for different sections. When compared with Table 2.2, the white matter is darker in T2 and the fluids are brighter. This modality is also one of the most widely considered MRI methods to study the pathological area in an image, which is quite brighter compared to other sections, as shown in Fig. 2.3.

Flair modality: This is another MRI variant that is widely adopted to record brain and spine examinations. Using flair, the signal obtained from the fluid is nullified by means of an extended successful echo time and extended inversion time. When it is used for brain/spine imaging, the lesions are covered with cerebrospinal fluid (CSF) signals. The flair modality MRI can be identified easily by verifying the contrast of the CSF (Table 2.4).

The flair MRI consists of three pixel groups, dark, gray, and bright, as depicted in Fig. 2.4. In this modality, the pathological area is more visible compared to T1 and T1C and is similar to T2-weighted images.

Diffusion-weighted modality: This is another MRI variant where the recording of the image depends on arbitrary microscopic movement of water protons, which is significantly modified by altered pathological procedures. DW MRI uses the T2-weighted pulse series with two additional slope pulses which

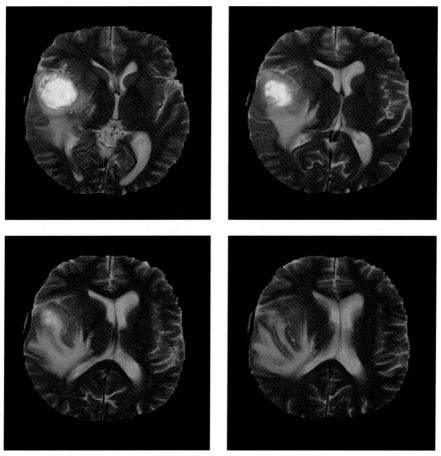

Figure 2.3 Example T2-weighted MRI of various areas.

Table 2.4 Appearance of various organ/tissue areas in a flair-weighted image.

Color	Area
Dark	Air
	Bone
	Blood flow
	White matter
	Fluids
Gray	Gray matter
	Muscles
Bright/white	Bone marrow
	Fat
	Pathological section

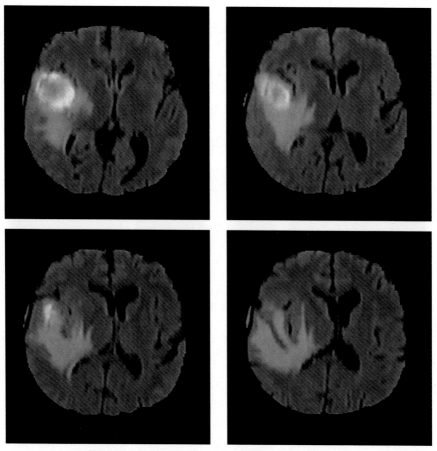

Figure 2.4 Example flair-weighted MRI of various areas.

are identical in amount and opposite in direction. These ramp pulses function by connecting the nuclear spin excitation and information acquisition. Hence, the attained image is very responsive to water molecule movement in the track of the other applied gradients. The additional gradient pulses lead to a reduction of the signal strength according to the water circulation in that area. A higher water distribution results in a greater signal reduction and the images are darker.

The example MRI of ISLES2015 is presented in Fig. 2.5. In this, the comparison of T1, flair, and DW is shown. The pathological area in T1 is less visible (dark/gray) compared to the other sections. In flair and DW image, the pathological areas are very bright in the considered image. In some images, the information available in the DW one is much less and most of the areas are not

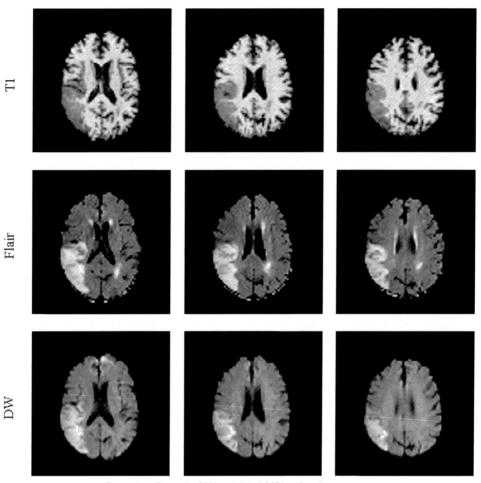

Figure 2.5 Example DW-weighted MRIs of various areas.

reconstructed properly as illustrated in Refs. [38,39]. The choice of modality depends on the disease and must be decided by the doctor after a discussion with the radiologist.

2.3 Verification and validation

Fig. 2.6 illustrates the traditional MRI machine used to record 3D images using a chosen modality. The figures consists of three major areas, a magnet to generate a uniform magnetic field, a gradient coil, and the RF coil. The gradient coil outcome is controlled with the help of X-, Y-, and Z-axis amplifiers and the RF coil outcome is controlled by an RF amplifier. Each operation is controlled with a scanning unit and a control panel associated with a computer/digital controller. When the recording procedure is initiated, the computer algorithm

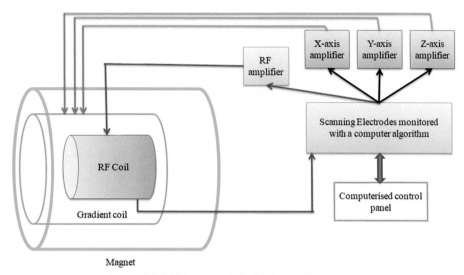

Figure 2.6 Architecture of the MRI recording system.

continuously provides the necessary control signal to the unit, to record the image based on the chosen procedure.

After recording the required image with this system, it is then verified by the radiologist and this verification helps to obtain the disease score. When the radiologist identifies any artifacts and unclear areas in the reconstructed image, they are corrected and the corrected image is then send to the doctor for further assessment and confirmation. Modern MRI scanning systems are equipped with a considerable number of facilities, which support fixing the scanning time, signal amplitude, duration of the scan, image reconstruction, and correction if necessary, with validation and report generation. Other information related to MRI recording can be accessed from Ref. [40]. The recorded images using this procedure are then verified by the doctor and based on the diagnostic report, and then the necessary treatment is planned and implemented.

2.4 Comparison of MRI modalities

Recording of the organ condition with a chosen MRI modality is widely adopted in the literature, and each modality has its own advantages and disadvantages. The choice of a particular modality depends on the body area to be evaluated, the patient's condition, and the expertise of the doctor. A comparison of different MRI modalities based on the pixel distribution is presented in Figs. 2.7 and 2.8 for BRATS and ISLES, respectively.

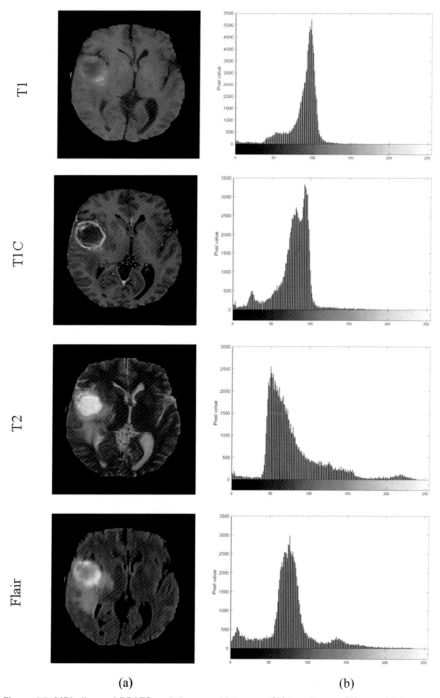

(a)   (b)

Figure 2.7 MRI slices of BRATS and the gray histogram: (A) test image; (B) gray histogram.

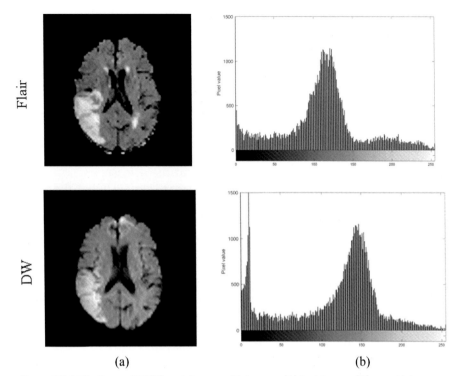

Figure 2.8 MRI slices of ISLES and the gray histogram: (A) test image; (B) gray histogram.

Fig. 2.7A presents an example test image and Fig. 2.7B shows the gray histogram for the chosen modality. The Y-axis of the histogram shows the pixel distribution and the X-axis presents its color, such as dark, gray, and bright. In T1 and T1C, there are more dark and gray pixels compared to the bright pixels, and the pixel distribution in these two cases is identical for the chosen image. The gray and bright pixels are more compared to the dark in both T2 and flair modality MRI. The pixel amplitude is less with T1C, T2, and flair than with T1. A similar comparison is presented in Fig. 2.8 using ISLES images. From Fig. 2.8A and B, it can be seen that the pixel value and threshold value are identical in the flair and DW modality images considered for this study.

A comparison based on the pixel distribution also helps to reveal the pixel groups in the chosen MRI slices, which separates the pathological area from the normal part using the chosen image-processing method [2,3,41,42].

2.5 Summary

Image-supported disease diagnosis is an approved medical procedures and a number of imaging methods are described in the literature to examine various abnormalities. MRI is one of the most commonly adopted imaging methodologies to detect disease in a number of internal organs. Compared to other approaches, MRI is still recommended by healthcare professionals due to its superiority and also its multimodality nature. Further, MRI supports the recording of a 3D image, which helps to provide complete information about the disease.

This chapter presented an overview about the various modalities (T1, T2, flair, and DW) used to record an abnormal body organ. Further, the information regarding the visibility of various areas based on a change in the modality has been also presented. An overview of the MRI recording procedure and its evaluation are also discussed briefly. From this discussion, it can be confirmed that MRI-supported disease diagnosis is one of the vital procedures in examining diseases in various body parts, such as the breasts, brain, spine, and heart.

References

[1] Marinaccio K, Kohn S, Parasuraman R, De Visser EJ. A framework for rebuilding trust in social automation across health-care domains. Proceedings of the international symposium on human factors and ergonomics in health care, vol. 4. New Delhi, India: SAGE Publications; June 2015. p. 201−5. 1.

[2] Dey N, Ashour AS, Beagum S, Pistola DS, Gospodinov M, Gospodinova EP, et al. Parameter optimization for local polynomial approximation based intersection confidence interval filter using genetic algorithm: an application for brain MRI image de-noising. Journal of Imaging 2015;1(1): 60−84.

[3] Tian Z, Dey N, Ashour AS, McCauley P, Shi F. Morphological segmenting and neighborhood pixel-based locality preserving projection on brain fMRI dataset for semantic feature extraction: an affective computing study. Neural Computing and Applications 2018;30(12):3733−48.

[4] Dey N, Rajinikanth V, Shi F, Tavares JMR, Moraru L, Karthik KA, et al. Social-Group-Optimization based tumor evaluation tool for clinical brain MRI of Flair/diffusion-weighted modality. Biocybernetics and Biomedical Engineering 2019;39(3):843−56.

[5] Kadry S, Rajinikanth V, Raja NSM, Hemanth DJ, Hannon NM, Raj ANJ. Evaluation of brain tumor using brain MRI with modified-moth-flame algorithm and Kapur's thresholding: a study. Evolutionary Intelligence 2021: 1−11.

[6] Tiwari A, Srivastava S, Pant M. Brain tumor segmentation and classification from magnetic resonance images: review of selected methods from 2014 to 2019. Pattern Recognition Letters 2020;131:244−60.

[7] Kadry S, Damaševičius R, Taniar D, Rajinikanth V, Lawal IA. Extraction of tumour in breast MRI using joint thresholding and segmentation—A study. In: 2021 seventh international conference on bio signals, images, and instrumentation (ICBSII). IEEE; March 2021. p. 1—5.

[8] Rajinikanth V, Priya E, Lin H, Lin F. Hybrid image processing methods for medical image examination. CRC Press; 2021.

[9] Pace DF, Dalca AV, Geva T, Powell AJ, Moghari MH, Golland P. Interactive whole-heart segmentation in congenital heart disease. In: International conference on medical image computing and computer-assisted intervention. Cham: Springer; October 2015. p. 80—8.

[10] Zhuang X, Shen J. Multi-scale patch and multi-modality atlases for whole heart segmentation of MRI. Medical Image Analysis 2016;31:77—87.

[11] Mukhopadhyay A. Total variation random forest: fully automatic mri segmentation in congenital heart diseases. In: Reconstruction, segmentation, and analysis of medical images. Cham: Springer; 2016. p. 165—71.

[12] Akhbardeh A, Jacobs MA. Comparative analysis of nonlinear dimensionality reduction techniques for breast MRI segmentation. Medical Physics 2012; 39(4):2275—89.

[13] Kang D, Shin SY, Sung CO, Kim JY, Pack JK, Choi HD. An improved method of breast MRI segmentation with Simplified K-means clustered images. In: Proceedings of the 2011 ACM symposium on research in applied computation; November 2011. p. 226—31.

[14] Fashandi H, Kuling G, Lu Y, Wu H, Martel AL. An investigation of the effect of fat suppression and dimensionality on the accuracy of breast MRI segmentation using U-nets. Medical Physics 2019;46(3):1230—44.

[15] Gubern-Merida A, Kallenberg M, Mann RM, Marti R, Karssemeijer N. Breast segmentation and density estimation in breast MRI: a fully automatic framework. IEEE Journal of Biomedical and Health Informatics 2014;19(1): 349—57.

[16] El Adoui M, Mahmoudi SA, Larhmam MA, Benjelloun M. MRI breast tumor segmentation using different encoder and decoder CNN architectures. Computers 2019;8(3):52.

[17] Liu L, Wu FX, Wang J. Efficient multi-kernel DCNN with pixel dropout for stroke MRI segmentation. Neurocomputing 2019;350:117—27.

[18] Mitra J, Bourgeat P, Fripp J, Ghose S, Rose S, Salvado O, et al. Lesion segmentation from multimodal MRI using random forest following ischemic stroke. Neuroimage 2014;98:324—35.

[19] Clèrigues A, Valverde S, Bernal J, Freixenet J, Oliver A, Lladó X. Acute and sub-acute stroke lesion segmentation from multimodal MRI. Computer Methods and Programs in Biomedicine 2020;194:105521.

[20] Subudhi A, Sahoo S, Biswal P, Sabut S. Segmentation and classification of ischemic stroke using optimized features in brain MRI. Biomedical Engineering: Applications, Basis and Communications 2018;30(03):1850011.

[21] Subudhi A, Dash M, Sabut S. Automated segmentation and classification of brain stroke using expectation-maximization and random forest classifier. Biocybernetics and Biomedical Engineering 2020;40(1):277—89.

[22] Maier O, Schröder C, Forkert ND, Martinetz T, Handels H. Classifiers for ischemic stroke lesion segmentation: a comparison study. PLoS One 2015; 10(12):e0145118.

[23] Akkus Z, Galimzianova A, Hoogi A, Rubin DL, Erickson BJ. Deep learning for brain MRI segmentation: state of the art and future directions. Journal of Digital Imaging 2017;30(4):449—59.

[24] Gordillo N, Montseny E, Sobrevilla P. State of the art survey on MRI brain tumor segmentation. Magnetic Resonance Imaging 2013;31(8):1426–38.

[25] Pereira S, Pinto A, Alves V, Silva CA. Brain tumor segmentation using convolutional neural networks in MRI images. IEEE Transactions on Medical Imaging 2016;35(5):1240–51.

[26] Wadhwa A, Bhardwaj A, Verma VS. A review on brain tumor segmentation of MRI images. Magnetic Resonance Imaging 2019;61:247–59.

[27] Pereira S, Alves V, Silva CA. Adaptive feature recombination and recalibration for semantic segmentation: application to brain tumor segmentation in MRI. In: International conference on medical image computing and computer-assisted intervention. Cham: Springer; September 2018. p. 706–14.

[28] Recht MP, Block KT, Chandarana H, Friedland J, Mullholland T, Teahan D, et al. Optimization of MRI turnaround times through the use of dockable tables and innovative architectural design strategies. American Journal of Roentgenology 2019;212(4):855–8.

[29] Weidman EK, Dean KE, Rivera W, Loftus ML, Stokes TW, Min RJ. MRI safety: a report of current practice and advancements in patient preparation and screening. Clinical Imaging 2015;39(6):935–7.

[30] Sessler JL, Mody TD, Hemmi GW, Lynch V, Young SW, Miller RA. Gadolinium (III) texaphyrin: a novel MRI contrast agent. Journal of the American Chemical Society 1993;115(22):10368–9.

[31] Ferris N, Goergen S. Gadolinium contrast medium (MRI contrast agents). Inside Radiology website 2016. insideradiology. com. au/gadolinium-contrast-medium/. Updated November, 22.

[32] https://mrimaster.com/.

[33] Lin D, Rajinikanth V, Lin H. Hybrid image processing-based examination of 2D brain MRI slices to detect brain tumor/stroke section: a study. In: Signal and image processing techniques for the development of intelligent healthcare systems. Singapore: Springer; 2021. p. 29–49.

[34] Acharya UR, Fernandes SL, WeiKoh JE, Ciaccio EJ, Fabell MKM, Tanik UJ, et al. Automated detection of Alzheimer's disease using brain MRI images–a study with various feature extraction techniques. Journal of Medical Systems 2019;43(9):1–14.

[35] Menze BH, Jakab A, Bauer S, Kalpathy-Cramer J, Farahani K, Kirby J, et al. The multimodal brain tumor image segmentation benchmark (BRATS). IEEE Transactions on Medical Imaging 2014;34(10):1993–2024.

[36] Bakas S, Reyes M, Jakab A, Bauer S, Rempfler M, Crimi A, et al. Identifying the best machine learning algorithms for brain tumor segmentation, progression assessment, and overall survival prediction in the BRATS challenge. arXiv 2018;1811:02629. preprint arXiv.

[37] Parry A, Clare S, Jenkinson M, Smith S, Palace J, Matthews PM. MRI brain T1 relaxation time changes in MS patients increase over time in both the white matter and the cortex. Journal of Neuroimaging 2003;13(3):234–9.

[38] Maier O, Menze BH, von der Gablentz J, Häni L, Heinrich MP, Liebrand M, et al. ISLES 2015-A public evaluation benchmark for ischemic stroke lesion segmentation from multispectral MRI. Medical Image Analysis 2017;35: 250–69.

[39] Winzeck S, Hakim A, McKinley R, Pinto JA, Alves V, Silva C, et al. ISLES 2016 and 2017-benchmarking ischemic stroke lesion outcome prediction based on multispectral MRI. Frontiers in Neurology 2018;9:679.

[40] Mullinger K, Debener S, Coxon R, Bowtell R. Effects of simultaneous EEG recording on MRI data quality at 1.5, 3 and 7 tesla. International Journal of Psychophysiology 2008;67(3):178−88.

[41] Priya E, Rajinikanth V. Signal and image processing techniques for the development of intelligent healthcare systems. Springer; 2020.

[42] Rajinikanth V, Raja NSM, Dey N. A beginner's guide to multilevel image thresholding. CRC Press; 2020.

3

Image processing methods to enhance disease information in MRI slices

3.1 Introduction

Globally, disease investigation procedures adopted in hospitals involve: (1) special evaluation by an expert and (2) recording and evaluation of the disease and its symptoms with the chosen procedure.

Recently, bioimage-supported disease diagnosis has been widely adopted in hospitals to diagnose diseases to enable efficient treatment. Although a considerable number of imaging procedures are available, MRI is most widely adopted due to its high reputation, 3D nature, and flexibility in the choice of different modalities [1–4].

Compared to other imaging methods, MRI presents an RGB scaled image (which also has dark, gray, and bright pixel combinations) and it can be evaluated in RGB or grayscale. Examination of an MRI in 3D form is quite complex and needs complex computation. Hence, 3D to 2D conversion is preferred to separate a 3D image into axial, coronal, and sagittal 2D slices. In the literature, due to its simplicity, the axial view is widely preferred for assessment [5–7].

After obtaining the 2D slice using a suitable procedure, the other preprocessing methods, such as resizing, artifact removal, contrast enhancement, image correction, and thresholding can be employed to enhance the disease information in the considered 2D MRI. After enhancing the image appropriately, a chosen postprocessing method can be implemented to extract and evaluate the disease information [8–10].

The MRI that comes directly from the radiology center is called a raw/unprocessed image and, in most cases, it is associated with artifacts and noise, and these image abnormalities must be corrected before it is submitted to the doctor for further examination. The remaining section of this chapter presents

Magnetic Resonance Imaging: Recording, Reconstruction and Assessment. https://doi.org/10.1016/B978-0-12-823401-3.00002-X

various image preprocessing and postprocessing procedures widely employed and described in the literature to adjust the MRI slices before analyzing them using semi/fully automated computerized methods. In this work, only MRI slices of breast, heart, and brain are considered for the demonstration and similar procedures can be adopted for other MRI cases.

3.2 Improvement methods for MRI slices

After recording the MRI using a chosen modality, its state must be improved using different enhancement and preprocessing methods. Commonly adopted image enhancement and preprocessing procedures are presented in this section with the appropriate results described.

The various MRI image processing procedures are depicted in Fig. 3.1 and the most appropriate method is chosen by researchers based on the disease to be examined and the modality of the MRI. If a 3D examination algorithm is available, then the 3D to 2D conversion can be ignored. Most of the methods depicted in Fig. 3.1 are applicable only for 2D MRI slices.

3.2.1 Conversion of 3D MRI into a 2D slice

The MRI obtained with a scanning procedure helps to obtain a 3D image which can be converted into 2D slices using suitable software. From the literature, it can be seen that ITK-Snap is one of the widely adopted open-source software to achieve this task.

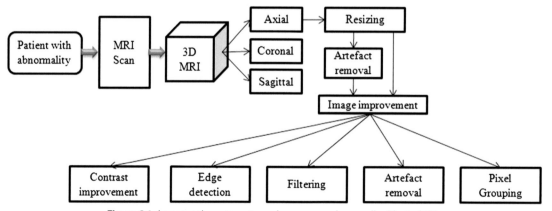

Figure 3.1 Image enhancements and preprocessing applicable to MRI.

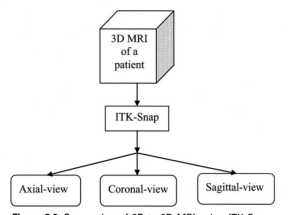

Figure 3.2 Conversion of 3D to 2D MRI using ITK-Snap.

Fig. 3.2 depicts the extraction of 2D slices using ITK-Snap [11,12] and Fig. 3.3 presents 2D slices of axial, coronal, and sagittal views attained from the MRI of varied modalities.

To demonstrate the need of the 3D to 2D conversion, in this sub-section, a number of benchmark images are considered and the extracted sample images are presented and discussed.

Fig. 3.3 presents an MRI and magnetic resonance angiography (MRA) recorded from a healthy volunteer. MRA is a variant of MRI used to study the activity/condition of blood vessels. The test images were obtained from Ref. [13]. This image is recorded without the contrast and hence the pixel levels are closer to gray and dark grades.

An MRI with varied modality is also considered in the literature to examine the brain abnormalities known as low-grade glioma (LGG) and high-grade glioma (HGG). Fig. 3.4 shows the LGG of the BRATS2015 data set [14]; the earlier works on this data set can be found in Refs. [15–17]. The LGG and HGG are widely considered brain abnormalities in the literature and they are commonly examined by researchers as tumors. The main merit of this data set is that it consists of T1, T2, T1C, and flair weighted images, and every image is associated with a related ground-truth provided by a disease expert. Fig. 3.5 shows the 2D slices of the HGG in the same data set and the related research with this data set can be found in Refs. [18–20].

Fig. 3.6 shoes 2D axial-view MRI slices obtained from the TCIA-GBM database [21,22]. This data set consists of the clinical-grade benchmark 3D MRI recorded with various modalities depicted in the figure. This database also consists of the functional MRI (fMRI) images, a modern form of MRI. Other information about

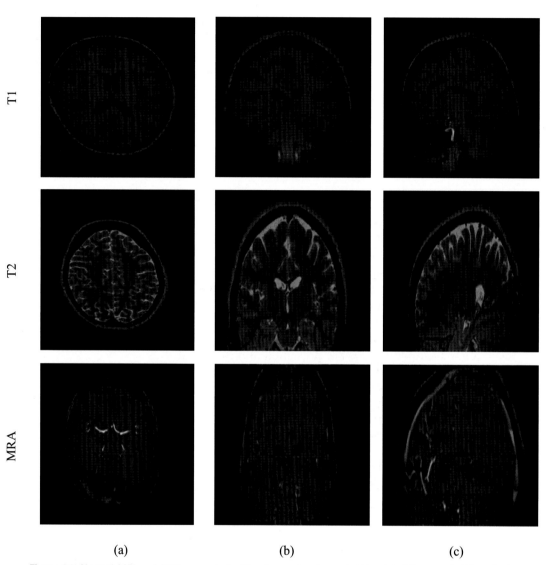

Figure 3.3 Normal MRI and MRA recorded without a contrast agent: (A) axial; (B) coronal; (C) sagittal.

the test images and their earlier research works can be accessed from Refs. [23—25].

Fig. 3.7 presents MRI slices of the multiple sclerosis (MS) brain MRI database. MS is one of the major abnormalities in humans caused by autoimmune sickness. The main cause of the disease is unpredictable and the lesions due to MS normally can be diagnosed using brain/spine images and, in most of cases, T2-weighted imaging is used. The presented test images are collected

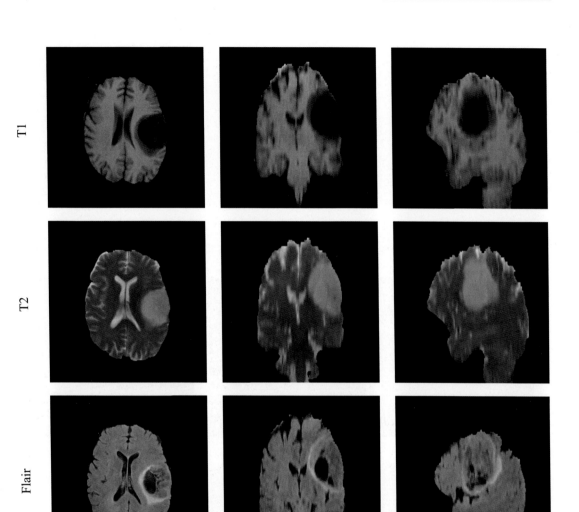

Figure 3.4 MRI slices of low-grade glioma: (A) axial; (B) coronal; (C) sagittal.

from the benchmark database and the earlier research on these can be found in Refs. [26,27].

Fig. 3.8 presents the 2D MRI slices of the schizophrenia images recorded using the T2 modality. This figure shows all three views extracted using the ITK-Snap, and these images were collected from the benchmark database available in Ref. [28]. Schizophrenia is also a common brain disorder in adults and early

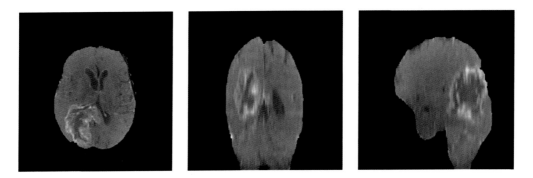

(a)   (b)   (c)

Figure 3.5 MRI slices of high-grade glioma: (A) axial; (B) coronal; (C) sagittal.

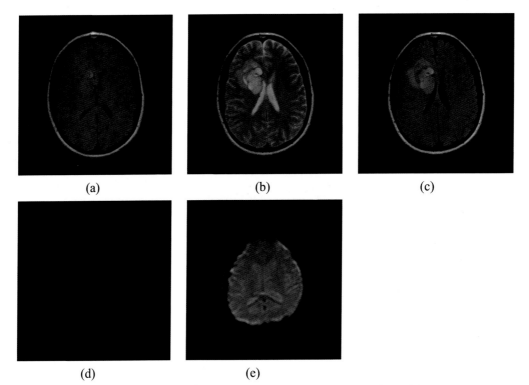
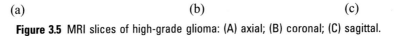

(a)   (b)   (c)

(d)   (e)

Figure 3.6 MRI slices of glioblastoma: (A) T1; (B) T2; (C) flair; (D) DW; (E) fMRI.

diagnosis can help to reduce the severity of the disease. A chosen machine-learning or deep-learning scheme is normally preferred to classify the MRI slices into normal and disease classes. The earlier research work implemented using this data set can be found in Ref. [29].

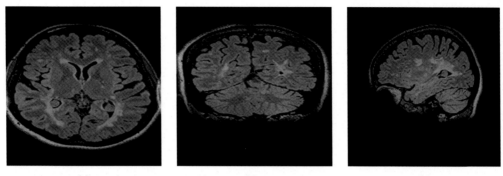

Figure 3.7 Multiple sclerosis brain MRI slides: (A) axial; (B) coronal; (C) sagittal.

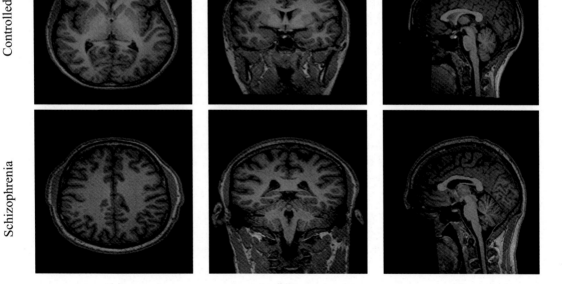

Figure 3.8 Healthy and schizophrenia class MRI slices: (A) axial; (B) coronal; (C) sagittal.

Fig. 3.9 shows 2D slices of the TCIA-RIDER database [30], which includes breast MRI slices along with the ground-truth. This is a clinical-grade MRI widely used by researchers to test their developed computer algorithms. Based on the need, the views, such as axial, coronal, and sagittal, can be considered for the

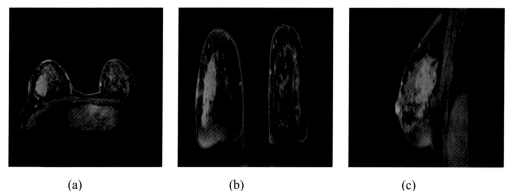

(a) (b) (c)

Figure 3.9 Breast MRI slices of the TCIA-RIDER database: (A) axial; (B) coronal; (C) sagittal.

assessment and the earlier research works carried with this database can be found in Refs. [31,32]. The earlier works also confirms that a breast cancer examined using MRI achieves better treatment results compared to mammograms, thermal images, and ultrasound.

Fig. 3.10 presents the breast MRI and MRA of the Radiopaedia data set [33]. In the MRI, the visibility of the tumor section is clear for both T1 and flair modality images. Further, the tumor and blood vessels enhanced with the contrast agent are very bright in the MRA. The earlier research work on the breast tumor diagnosis with these images can be found in Ref. [10].

Similar to the brain and breast, abnormalities in the heart also are normally examined with the heart MRI and the sample test images obtained from the HVSMR 2016 [34]. Mining of the blood pool and myocardium from a 3D cardiovascular MRI is a

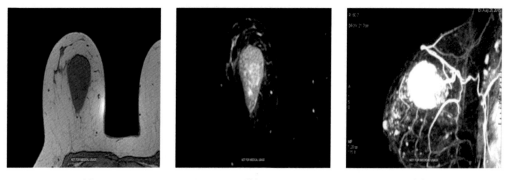

(a) (b) (c)

Figure 3.10 Breast MRI (axial) and MRA (sagittal) of the abnormal category: (A) T1; (B) flair; (C) MRA.

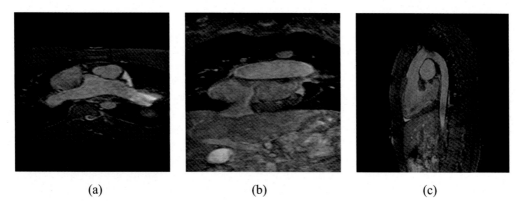

(a)   (b)   (c)

Figure 3.11 Heart MRI collected from HVSMR 2016: (A) axial; (B) coronal; (C) sagittal.

requirement prior to making a patient-specific heart examination for preprocedural arrangement of children with congenital inherited heart disease. The cardiovascular MRI is planned using a chosen modality and, after obtaining the image, the selected procedure is employed to extract the suspicious area for further evaluation. The earlier research work on the images shown in Fig. 3.11 can be accessed in Refs. [35,36].

In a variety of images recorded by means of the selected modality, pixel significance can be used to identify significant information in the images. In some situations, the information recorded in the untreated pictures is difficult to distinguish and hence a number of correction and processing actions are executed to enhance the raw MRI for the chosen modality and view. These schemes can help to enhance the information in raw MRI slices, and some of these procedures include contrast improvement, edge recognition, filtering, multimodality fusion, thresholding, segmentation, etc. The earlier works also confirmed that the complexity of the image examination will change based on the dimensions, therefore it is necessary to choose the dimensions of the image for computerized processing. In this research work, every image is resized to $256 \times 256 \times 1$ for a general image processing task and a smaller dimension for the deep-learning task.

3.2.2 Image resizing

Image resizing is a preliminary work in the image processing application and is widely adopted to resize test images. The earlier works in the literature confirm that the complexity and examination time of the image processing task will increase with its pixel dimension. The shorter image will be processed easily (a lower number of pixels to be inspected) and the larger image needs a longer processing time when computerized disease diagnosis is employed.

Fig. 3.12 presents the variation in the image pixel distribution based on the dimension. In this work, the commonly adopted image dimensions are considered for the assessment. Various image dimensions shown in Fig. 3.12A are considered for the assessment and the related image and histograms are shown in Fig. 3.12B and C, respectively. Fig. 3.12C confirms that, based on the change in image dimension, the pixel level of the image changes. However, the threshold value (histogram) looks the same for all the images depicted in this figure.

This confirms that the correct image dimension must be selected during the computerized image examination and the essential information regarding the dimension selection can be obtained from prior knowledge or from the literature. This concept is similar for the RGB scale images, and this information can be found in Refs. [37,38].

3.2.3 Artifact removal

This process is optional in real clinical examinations and is adopted only in computerized MRI assessment procedure,. Usually, the obtained MRI slice using 3D to 2D conversion is associated with various sections along with the disease-affected region which is to be examined. The unwanted sections in the MRI are referred to as artifacts, which have no information regarding the disease to be diagnosed. For example, if we consider a brain MRI slice, the skull section can be removed during the computerized examination task, since it will not provide any useful data about the disease. The earlier works confirm that the skull region can be extracted using a carefully tuned threshold-filter [39,40]. When this scheme is employed, it will separate the image into skull and tissue regions, and then the tissue section can be considered for further assessment.

Fig. 3.13 shows the scheme employed to separate a brain MRI slice into the skull and tissue sections. Fig. 3.14 presents an image and histogram of the original brain, skull, and tissue images and confirms that the combination of a tissue and skull histogram is similar to the histogram of the brain.

The chief merit of the threshold filter is that it separates the image section from the artifact and this process reduces the complexity during MRI slice assessment using the selected computer algorithm. The main limitation in threshold-filter

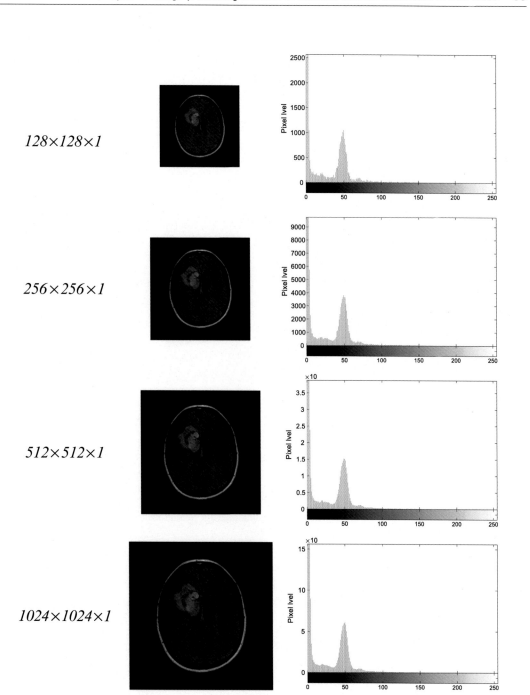

Figure 3.12 Change in pixel distribution based on the image dimension: (A) dimension; (B) test image; (C) histogram.

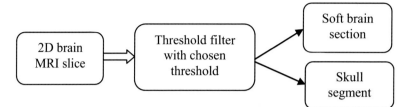

Figure 3.13 Skull stripping using a threshold filter.

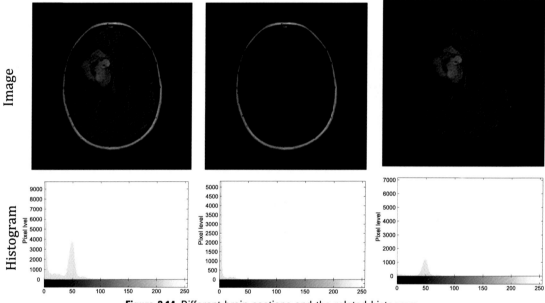

Figure 3.14 Different brain sections and the related histogram.

implementation is the selection of the finest threshold, which separates the image into two sections. Usually, the threshold selection is completed manually after a variety of trials. Experimentation process-based threshold selection is a time-consuming work.

3.2.4 Image correction

The MRI is recorded in a controlled location using the appropriate scanning procedure. During this process, the patient is requested not to move their body until the recording procedure is completed. During the recording, if the patient moves their body, the image recorded will be distorted. The image recording will not be repeated again, and hence, the image correction scheme is necessary to fix the abnormality in the recorded image.

In the literature, a considerable number of reconstruction and distortion removal procedures are proposed and implemented to obtain better clarity of the MRI slice. Fig. 3.15 presents the sample test images considered for this study. It presents both the distorted and reconstructed images of chosen MRI slides. The related information about the methodology employed can be found in Ref. [41]. The distortion removal employs a chosen methodology (distortion filter), which helps to correct the pixel distribution in the image to obtain the corrected image as shown in Fig. 3.16.

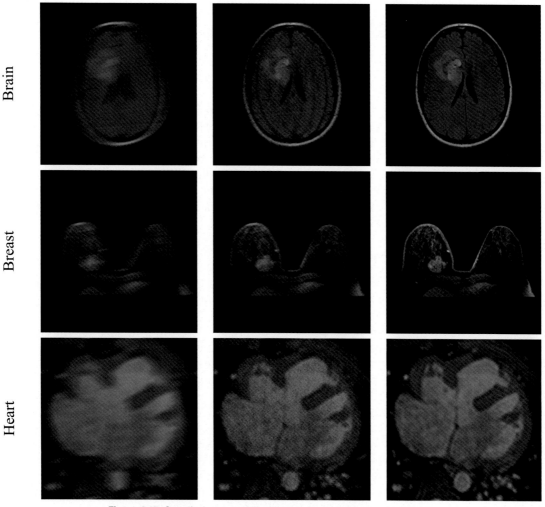

Figure 3.15 Sample images of the distorted and reconstructed image.

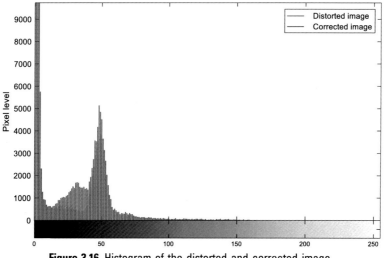

Figure 3.16 Histogram of the distorted and corrected image.

The image filter-supported pixel adjustment corrects the excess pixels in the selected MRI slice and is preferred as a suitable preprocessing scheme during the computerized examination of the selected images recorded using varied modalities. The chief limitation is the selection and retuning procedure of the filter is time consuming and, in most of the cases, the currently available image processing scheme can work well even if the MRI is distorted.

3.2.5 Noise removal

While recording using a chosen procedure, noise can occur for a variety of reasons, including external disturbances. This noise will degrade the information (image pixels) considerably and it must be corrected before sending the MRI slice for the doctor's attention. In order to demonstrate the noise removal performance in various MRI images, every image is corrupted using traditional salt-and-pepper noise and then it is removed using the noise filter with a chosen value. Every outcome stage of this process is depicted in Fig. 3.17, and a description of this process can be found in Refs. [42,43].

Noise filter-supported pixel regulation removes the highly visible pixels from MRI slice sand implementation of the noise filter needs to be tuned based on the intensity to be removed. The principal restriction in the filter is that the choice and retuning procedure of the filter is time consuming.

Brain

Breast

Heart

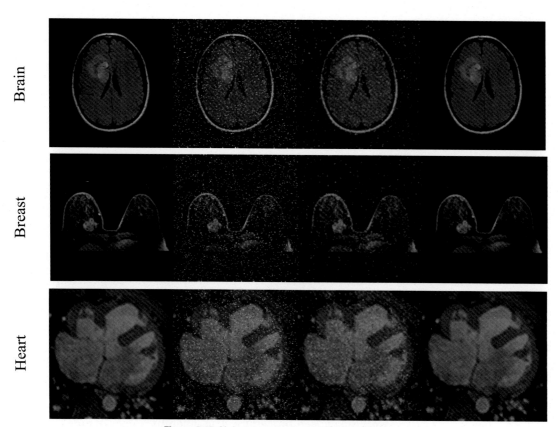

Figure 3.17 Noise-corrupted and corrected image.

3.2.6 Pixel-level correction

Normally the data seen in a grayscale image are relatively reduced compared to an RGB image and hence pixel-level correction is sometimes necessary. MRI is in grayscale form and improving the suspicious section's intensity with respect to the background is a difficult process. Therefore a number of image enhancement procedures have been proposed by researchers to improve the visibility of the test MRI slice. Contrast improvement is a commonly employed technique in computerized medical image assessment and this can be realized using a number of techniques. Procedures, like histogram equalization, color map tuning, and contrast-limited adaptive histogram equalization (CLAHE) are widely used pixel-level enhancement techniques employed to improve the intensity of grayscale images [44].

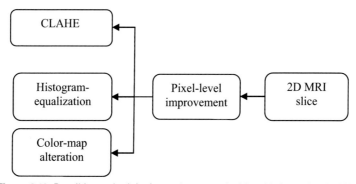

Figure 3.18 Possible methodologies to improve pixel-level information in MRI.

Fig. 3.18 shows a possible procedure to improve the pixel information, and its practical realization is shown in Fig. 3.19. From this outcome, it is observed that the employed methods will assist in improving the visibility of abnormal section contrast to the raw MRI slice. An investigational study using the chosen MRI slices of brain, breast, and heart confirmed that this procedure enhanced the visibility of suspicious areas considerably, and it can be easily recognized by employing the computerized examination technique.

The pixel enhancement scheme applied to raw MRI helps to improve the tumor segment by adjusting its pixel distribution with the chosen procedure. The different shapes of gray level histograms of the MRI slices with the implemented technique are depicted in Fig. 3.20. The selection of the pixel modification method depends on the needs and knowledge of the operator who implements the process for a chosen image. More information about this method can be found in Ref. [45].

3.2.7 Edge detection

Edge detection (ED) is widely used to detect the image borders of the area of interest and other regions in an MRI slice and, in the literature, a number of edge detection procedures have been described, with one of the main ones being the well-known image-mining technique called the watershed algorithm. The Canny-based ED is a commonly used method developed in 1986 [46], which uses a multistage algorithm to distinguish image edges. Canny ED uses valuable structural data from different

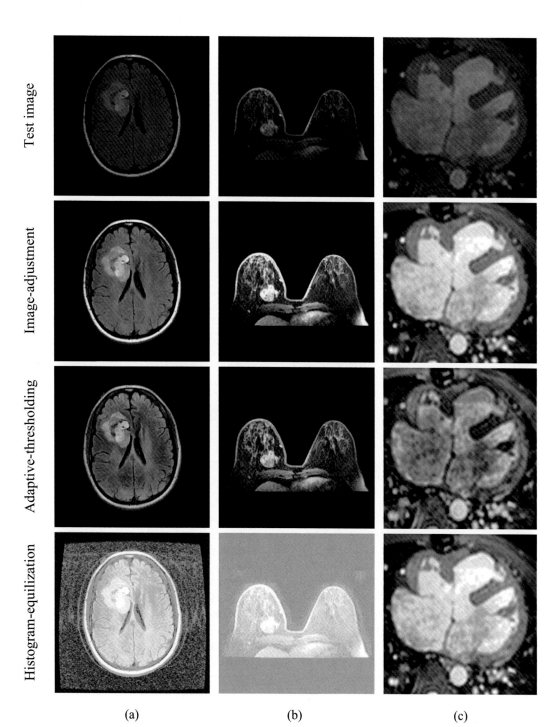

Figure 3.19 Enhanced MRI slices using various pixel adjustment procedures: (A) brain; (B) breast; (C) heart.

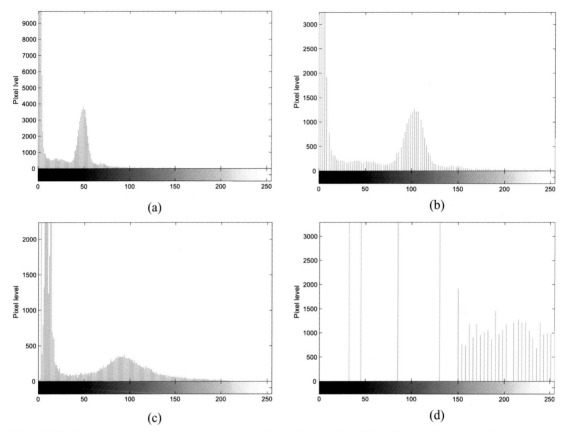

Figure 3.20 Histogram of various images enhanced with pixel correction: (A) test image; (B) image adjustment; (C) adaptive thresholding; (D) histogram equalization.

images and significantly reduces the amount of information to be inspected. The common conditions for ED include:

- Detection of the border with a reduced error value using precise location of all the edges existing in the MRI slice;
- The boundary angle expected from the operator must precisely confine the edges formed with the scheme;
- The border of the MRI must only be observed once, and it must be free from false boundaries.

To accomplish these requirements, Canny executed calculus of variations for optimization. Sobel is also one of the general edge recognition process categories employed to process images [47,48]. The ED outcomes achieved with Sobel and Canny are depicted in Fig. 3.21. Based on the requirements, either scheme is employed to extract the useful information from the MRI slice under test.

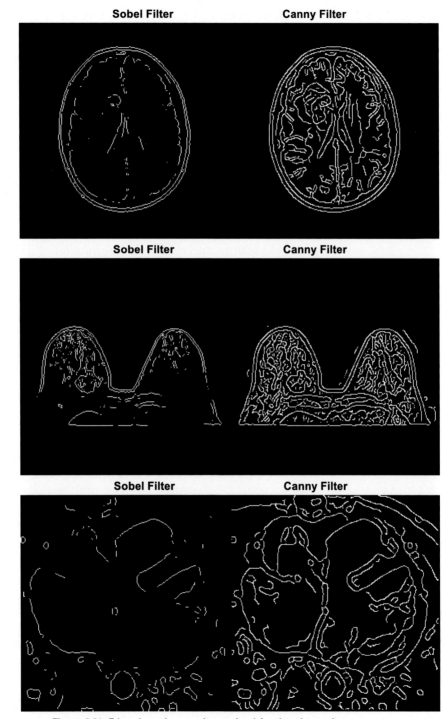

Figure 3.21 Edge detection results attained for the chosen images.

The advantage of edge detection is that it helps to identify the essential area in the image to be examined and helps to obtain the complete boundary and texture of the picture to be processed. A limitation of ED is that in some image cases (MRI slice with poor visibility), it will not provide the required detection and this process works only for the grayscale version of the image.

3.2.8 Gaussian filter-based enhancement

Computerized disease detection and classification is extensively considered in the literature for automated disease detection with a chosen digital bioimage. The essential information regarding Gaussian filter (GF)-based disease detection can be found in Refs. [49,50]. This procedure helps to obtain the improved texture and silhouette information of the image under study and the features extracted with a chosen procedure will help to classify the image with better accuracy. Previous work with GF confirmed that it can be employed to improve the surface and border features of a grayscale image of a chosen pixel value [51]. Further, one of the earlier research works investigated that the GF with varied scale (ϕ) presents an improvement of the texture pattern in vertical/horizontal directions, which presents the essential details to support the automated diagnosis.

Fig. 3.22 shows the experimental results with the chosen brain/breast MRI slice for $\phi = 60$ and $\phi = 300$. After the improvement, the exterior and border features can be extracted to develop the appropriate disease detection method.

GF is considered to normalize the exterior and edge of the grayscale MRI slice of a chosen organ. This idea can be considered to generate a variety of edge and texture patterns with respect to the angle ϕ. The features found in GF-treated MRI can be inspected only using the chosen methodology and it is only applicable to the automated imaging schemes.

3.2.9 Local binary pattern

Local binary pattern (LBP)-supported image enhancement is applicable only to the grayscale image and this procedure is widely adopted in automated disease diagnosis schemes. The earlier works on the LBP method can be found in Ref. [52], and this technique helps to get a 1D LBP feature of dimensions $1 \times 1 \times 59$ for every image treated using the LBP with a preferred weight (W). The work of Gudigar et al. [53] confirmed that the LBP with varied weights (W = 1, 2, 3, and 4) helps to provide better disease

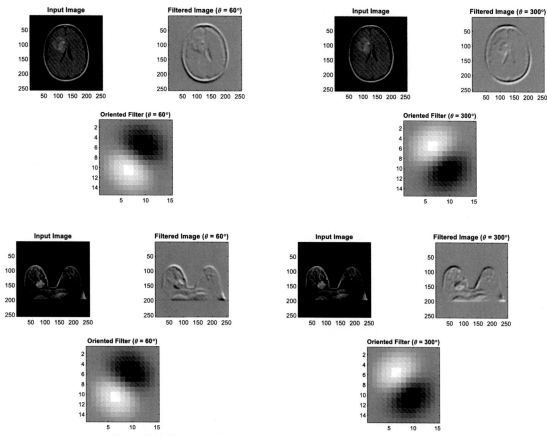

Figure 3.22 Texture enhancement achieved with MRI slices using GF.

detection compared to other related features-assisted disease detection procedures reviewed in the literature. In LBP, every pixel is grouped to its related pixel to form a new image which provides the essential information about the abnormality in the image. The LBPs obtained with brain, breast, and heart MRIs with a chosen weight are presented in Fig. 3.23 and a complete discussion regarding the LBP can be found in Ref. [54]. Fig. 3.23A and B depicts the LBP image and threshold for weight W = 1−4 respectively, and the change in the pattern based on this weight can also be seen clearly in these images. The extracted features are then considered to train and validate the classifier employed in the automated disease detection scheme.

The main advantage of this approach is that it will improve the image's texture and edge information and helps to get necessary features from the images with reduced complexity.

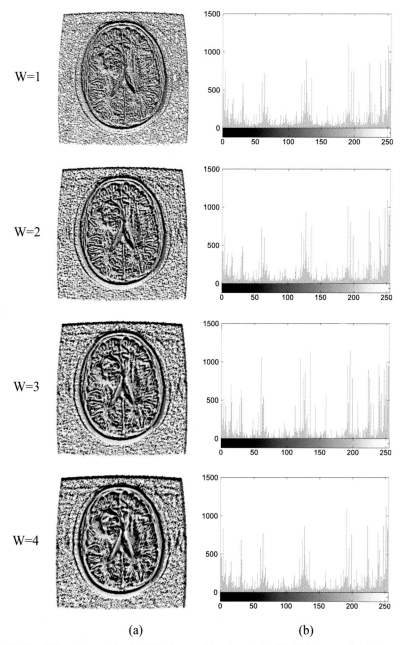

W=1

W=2

W=3

W=4

(a)

(b)

Figure 3.23 The LBP-enhanced brain MRI slice and its threshold: (A) LBP image; (B) LBP threshold.

3.2.10 Saliency-based enhancement

Saliency based enhancement helps to improve the image pixels, which appear abnormal compared to other normal pixels. The saliency technique is generally used to identify the pixel group, which provides the vital information regarding the disease in the MRI slice. Saliency detection can be applied for gray- and RGB-scaled images and, in this work, the grayscale MRI slices are treated with this technique. The essential information about the saliency detection in bioimages can be found in Refs. [55,56].

Figs. 3.24 and 3.25 present the saliency-enhanced brain MRI slice (depicted in Fig. 3.12) and the 3D feature map, respectively. Fig. 3.24 presents the hot-color map of the attained result, in which the prime pixels (skull and tumor) are more visible compared to other brain parts. When a skull-stripped image is used, this section helps to obtain the information about the abnormality existing in the brain MRI. The features available in the feature map are then used to develop automated disease detection methods which detect/classify the disease using the MRI slices. The saliency combined with the morphological segmentation helps to identify the abnormal section in the MRI.

3.2.11 Recent advancements in MRI enhancement

The image processing literature confirms that manual examination for a disease is time-consuming during a mass screening process (when a larger number of patients are tested) and to reduce the diagnostic burden on healthcare workers, a number

Saliency Map

Mean Feature Map

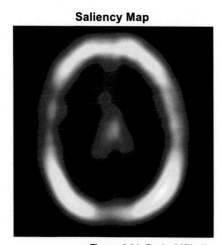 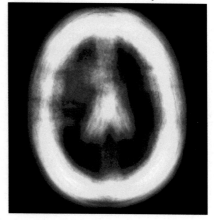

Figure 3.24 Brain MRI slice enhanced using the saliency process.

Figure 3.25 Saliency feature map.

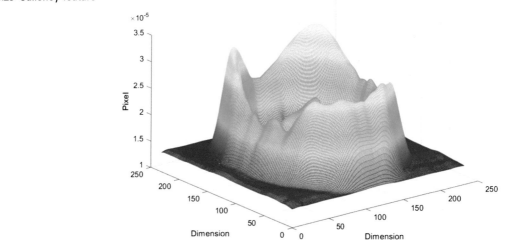

of disease detection methods are used. The examination of disease with a biomedical method is an approved methodology and hence a number of segmentation, machine-learning (ML), and deep-learning (DL) methods are proposed to diagnose a variety of diseases using bioimages.

Examination of MRI slices with traditional and convolutional neural network (CNN)-based methods are widely applied to detect and classify images.

In the literature, the possible combination of multithresholding and segmentation is commonly employed to extract the abnormal section in the chosen MRI slice. After the extraction, a comparison of the mined abnormal region is compared with the GT and based on the attained quality measures (QM), the advantage of the implemented scheme is confirmed. CNN-based segmentation is also widely employed to extract the suspicious parts of an MRI as discussed in Refs. [57–59]. The employment of feature-based ML and DL classification is also widely employed to classify MRI slices.

Because of their advantages, automated methods are implemented to detect abnormal areas in brain, breast, and heart MRIs. The recent MRI examination methods include: (1) heuristic algorithm-based preprocessing; (2) heuristic method-based feature selection; (3) a hybrid scheme; and (4) dual-deep architectures to examine the MRI. The choice of method depends on the disease to be examined, modality of the MRI, and the number of MRI slices available for the examination. When the MRI dimensions are large, then the DL scheme will be preferred, and for small and medium-sized MRI slices, the ML scheme is implemented.

3.2.12 Hybrid image assessment scheme

The medical image examination literature verifies the necessity of the hybrid image assessment procedure as this method helps to obtain better disease detection compared with the traditions methods.

Fig. 3.26 presents the heuristic algorithm-assisted joint thresholding segmentation scheme employed to extract the suspicious area in an MRI. The MRI slice to be examined is initially enhanced using the multithresholding operation using a chosen procedure (Otsu/entropy). The role of the heuristic algorithm is to select the optimal threshold based on the assigned threshold by the operator. After image enhancement, the required area is segmented and compared with the GT to compute the necessary QM. Based on the computed QM, such as Jaccard index (JI), dice (DC), accuracy (ACC), precision (PRE), sensitivity (SEN), specificity (SPE), and negative predictive value (NPV), the eminence of the proposed algorithm is confirmed and the segmented section is examined by the doctor and, based on the outcome, the treatment plan is initiated.

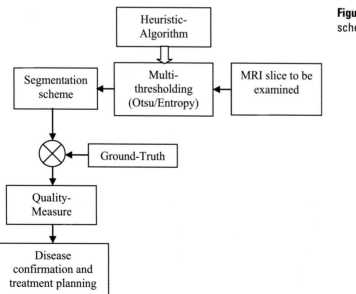

Figure 3.26 Hybrid image examination scheme.

Multithresholding based on the chosen scheme is widely executed in a range of areas to preprocess the test image. An image can be recognized as a collection of dissimilar pixels with respect to thresholds. In a digital image, the allocation of pixels plays a lead role. The regulation or grouping of image pixels is widely adopted to improve/adjust the information available in the image.

Bilevel thresholding is chosen to divide an unprocessed image into the suspicious area and the background. During this task, the expert/computer algorithm is permitted to identify the finest threshold with the help of the chosen method. Let a specific image have 256 thresholds (ranging from 0 to 255) and the role of the expert/computer algorithm is to discover the threshold (Th) which separates the image, as shown in Eq. (3.1):

$$0 < Th < 255 \tag{3.1}$$

The chosen threshold will support splitting the image into two segments; such as phase 1 with pixel allocation $< Th$ and phase 2 with pixel allocation $> Th$ during the bilevel thresholding.

If the MRI slice is to be separated into three areas, such as the suspicious area, normal part, and background, then trilevel thresholding is implemented, and this will help to obtain two thresholds ($Th1$ and $Th2$) which lie between the minimum (0) and maximum (255) thresholds in the image.

In a trilevel threshold, Eq. (3.2) is applicable:

$$0 < Th_1 < Th_2 < 255 \tag{3.2}$$

When Eq. (3.2) is executed, the image is divided into three sections: (1) pixels between 0 and Th_1, (2) pixels between Th_1 and Th_2, and (3) pixels between Th_2 and 255. In the MRI examination, the outcome of the bilevel threshold is not suitable and hence trilevel thresholding is widely employed [2,10,31].

Implementation of a suitable method to enhance the MRI is a difficult task and the appropriate MRI assessment procedures are chosen based on suggestions from previous works or experience. There are a number of thresholding schemes, and the generally implemented methods are shown in Fig. 3.27. Each approach has its own qualities and, in the literature, the histogram-supported method discussed above is widely recommended for preprocessing of medical images.

In histogram-supported thresholding, an examination of the histogram with an adopted technique is carried out to select the best based on a random search. During this task, a preferred objective value (entropy/Otsu between class variance) is

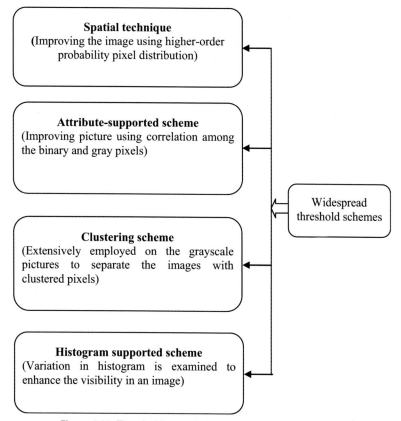

Figure 3.27 Thresholding technique for image enhancement.

considered as the guiding means to validate the excellence of the outcome based on the chosen QM.

Thresholding implemented for grayscale MRI works well for RGB scale MRI also with the only difference being that, in RGB MRI, the necessary threshold is identified by considering the R, G, and B channel histograms, which is quite complex compared to the grayscale MRI.

The outcome of Kapur-based thresholding for $Th = 2-5$ is presented in Fig. 3.28 for a selected brain MRI image. This figure depicts the image result and the corresponding threshold value. A similar procedure was implemented with Otsu and the attained results are presented in Fig. 3.29.

The earlier work on image thresholding-based MRI assessment confirmed that the entropy-supported scheme presents a better outcome as compared to the Otsu and hence, a number

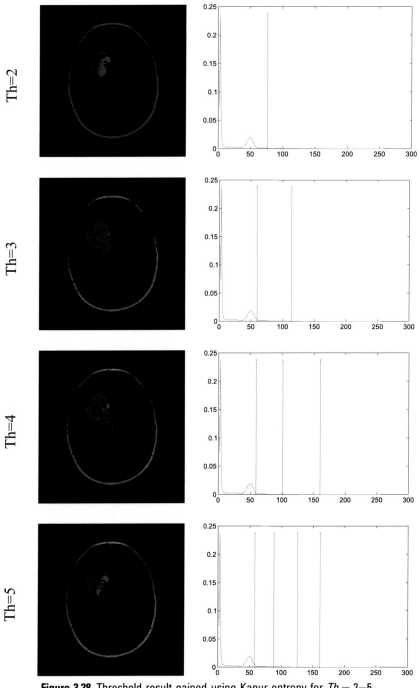

Figure 3.28 Threshold result gained using Kapur entropy for $Th = 2-5$.

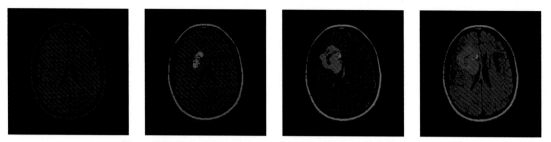

Figure 3.29 Threshold outcome achieved with Otsu for $Th = 2-5$.

of biomedical images are preprocessed using various entropy-based functions. The suspicious area from this enhanced image can be extracted using the chosen segmentation procedure. The essential features from these images are then extracted to develop the computerized image classification system which groups the considered MRIs into normal and abnormal classes.

The advantage of the thresholded result can be evaluated using a relative investigation of a raw image (I) and threshold image (T). In this process, every pixel of the image is individually evaluated and the result is used to discover the vital image quality parameters. The superiority of T is confirmed using image metrics, such as the toot mean squared error (RMSE), peak signal-to-noise ratio (PSNR), mean structural similarity index (MSSIM), normalized absolute error (NAE), normalized cross-correlation (NCC), average difference (AD), and structural content (SC), and these values are presented in Eqs. (3.3)−(3.9).

$$PSNR_{(I,T)} = 20 \log_{10}\left(\frac{255}{\sqrt{MSE_{(I,T)}}}\right); \text{dB} \qquad (3.3)$$

$$RMSE_{(I,T)} = \sqrt{MSE_{(I,T)}} = \sqrt{\frac{1}{XY}\sum_{i=1}^{X}\sum_{j=1}^{Y}\left[I_{(i,j)} - T_{(i,j)}\right]^2} \qquad (3.4)$$

MSSIM is usually computed to approximate the picture supremacy and interdependencies among the original and threshold images.

$$MSSIM_{(I,T)} = \frac{1}{M}\sum_{z=1}^{M} SSIM_{(I_z,T_z)} \qquad (3.5)$$

where I_z and T_z are the image contents at the z-th local window; and M is sum of the local windows of the image.

$$\text{NAE}_{(I,T)} = \frac{\sum\limits_{i=1}^{X}\sum\limits_{j=1}^{Y}\left| I_{(i,j)} - T_{(i,j)} \right|}{\sum\limits_{i=1}^{X}\sum\limits_{j=1}^{Y}\left| I_{(i,j)} \right|} \qquad (3.6)$$

$$\text{NCC}_{(I,T)} = \frac{\sum\limits_{i=1}^{X}\sum\limits_{j=1}^{Y} I_{(i,j)} \cdot T_{(i,j)}}{\sum\limits_{i=1}^{X}\sum\limits_{j=1}^{Y} I_{(i,j)}^2} \qquad (3.7)$$

$$\text{AD}_{(I,T)} = \frac{\sum\limits_{i=1}^{X}\sum\limits_{j=1}^{Y} I_{(i,j)} - T_{(i,j)}}{XY} \qquad (3.8)$$

$$\text{SC}_{(I,T)} = \frac{\sum\limits_{i=1}^{X}\sum\limits_{j=1}^{Y} I_{(i,j)}^2}{\sum\limits_{i=1}^{X}\sum\limits_{j=1}^{Y} T_{(i,j)}^2} \qquad (3.9)$$

In all these equations, $X \times Y$ is the size of the image, R is the original test image, and S is the segmented image of a chosen threshold. Higher values of PSNR, MSSIM, and NCC, and lower values of RMSE, NAE, AD, and SC indicate a superior quality of thresholding.

3.3 Summary

MRI-based disease detection is widely adopted in hospitals to detect abnormal areas from 3D and 2D images. The assessment of a 3D image is quite complex and hence 3D to 2D conversion is widely employed during the disease diagnosis. In brain MRI, the recorded image is associated with the skull section and hence a skull stripping process is normally executed with a chosen threshold filter. Further, in order to obtain an MRI slice with better visibility, a number of procedures, such as resizing, correction, filtering, and contrast enhancement methods are widely employed by researchers to get enhanced results during the MRI examination using the proposed algorithm. Along with these methods, the thresholding-assisted MRI examination is also used to enhance the visibility of the suspicious area based on the pixel grouping concept. In this chapter, Kapur- and Otsu-based thresholding with a chosen threshold (*Th*) value of 25 is demonstrated using a brain MRI with tumor, and the obtained results confirmed that the Kapur technique presented with better results that the Otsu method. Based on the requirements, an examination using MRI can be implemented using the methods discussed in this chapter. All the methods discussed herein are applicable for MRI with varied modalities and varied views, including axial, coronal, and sagittal.

References

[1] Van Den Brekel MW. Lymph node metastases: CT and MRI. European Journal of Radiology 2000;33(3):230–8.

[2] Abirami D, Shalini N, Rajinikanth V, Lin H, Rao VS. Brain MRI examination with varied modality fusion and Chan-Vese segmentation. In: Intelligent data engineering and analytics. Singapore: Springer; 2021. p. 671–9.

[3] Tian Z, Dey N, Ashour AS, McCauley P, Shi F. Morphological segmenting and neighborhood pixel-based locality preserving projection on brain fMRI dataset for semantic feature extraction: an affective computing study. Neural Computing and Applications 2018;30(12):3733–48.

[4] https://mrimaster.com/.

[5] Arunmozhi S, Raja NSM, Rajinikanth V, Aparna K, Vallinayagam V. Schizophrenia detection using brain MRI—a study with watershed algorithm. In: 2020 international conference on system, computation, automation and networking (ICSCAN). IEEE; July 2020. p. 1–4.

[6] Shi F, Dey N, Ashour AS, Sifaki-Pistolla D, Sherratt RS. Meta-KANSEI modeling with valence-arousal fMRI dataset of brain. Cognitive Computation 2019;11(2):227–40.

[7] Mbarki W, Bouchouicha M, Tshienda FT, Moreau E, Sayadi M. Herniated lumbar disc generation and classification using cycle generative adversarial networks on axial view MRI. Electronics 2021;10(8):982.

[8] Chen Y, Chen G, Wang Y, Dey N, Sherratt RS, Shi F. A distance regularized level-set evolution model based MRI dataset segmentation of brain's caudate nucleus. IEEE Access 2019;7:124128–40.

[9] Dey N, Ashour AS, Beagum S, Pistola DS, Gospodinov M, Gospodinova EP, et al. Parameter optimization for local polynomial approximation based intersection confidence interval filter using genetic algorithm: an application for brain MRI image de-noising. Journal of Imaging 2015;1(1): 60–84.

[10] Raja NSM, Fernandes SL, Dey N, Satapathy SC, Rajinikanth V. Contrast enhanced medical MRI evaluation using Tsallis entropy and region growing segmentation. Journal of Ambient Intelligence and Humanized Computing 2018:1–12.

[11] Yushkevich PA, Piven J, Hazlett HC, Smith RG, Ho S, Gee JC, et al. User-guided 3D active contour segmentation of anatomical structures: significantly improved efficiency and reliability. Neuroimage 2006;31(3): 1116–28.

[12] http://www.itksnap.org/pmwiki/pmwiki.php.

[13] Bullitt E, Smith J, Lin W. Designed database of MR brain images of healthy volunteers. MIDAS Insight-Journal 2011. http://insight-journal.org/midas/community/view/21.

[14] Menze BH, Jakab A, Bauer S, Kalpathy-Cramer J, Farahani K, Kirby J, et al. The multimodal brain tumor image segmentation benchmark (BRATS). IEEE Transactions on Medical Imaging 2014;34(10):1993–2024.

[15] Rao V, Sarabi MS, Jaiswal A. Brain tumor segmentation with deep learning. MICCAI Multimodal Brain Tumor Segmentation Challenge (BraTS) 2015;59.

[16] Saba T, Mohamed AS, El-Affendi M, Amin J, Sharif M. Brain tumor detection using fusion of hand crafted and deep learning features. Cognitive Systems Research 2020;59:221−30.

[17] Li H, Li A, Wang M. A novel end-to-end brain tumor segmentation method using improved fully convolutional networks. Computers in Biology and Medicine 2019;108:150−60.

[18] Cui S, Mao L, Jiang J, Liu C, Xiong S. Automatic semantic segmentation of brain gliomas from MRI images using a deep cascaded neural network. Journal of Healthcare Engineering 2018;2018.

[19] Monteiro M, Figueiredo MA, Oliveira AL. Conditional random fields as recurrent neural networks for 3d medical imaging segmentation. arXiv 2018; 1807:07464. preprint arXiv.

[20] Cho HH, Park H. Classification of low-grade and high-grade glioma using multi-modal image radiomics features. In: 2017 39th annual international conference of the IEEE engineering in medicine and biology society (EMBC). IEEE; July 2017. p. 3081−4.

[21] Clark K, Vendt B, Smith K, Freymann J, Kirby J, Koppel P, et al. The Cancer Imaging Archive (TCIA): maintaining and operating a public information repository. Journal of Digital Imaging 2013;26(6):1045−57.

[22] Scarpace L, Mikkelsen T, Cha S, Rao S, Tekchandani S, Gutman D, et al. Radiology data from the cancer genome atlas glioblastoma multiforme. The Cancer Imaging Archive 2016. https://doi.org/10.7937/K9/ TCIA.2016.RNYFUYE9 [TCGA-GBM] collection [Data set].

[23] Kadry S, Rajinikanth V, Raja NSM, Hemanth DJ, Hannon NM, Raj ANJ. Evaluation of brain tumor using brain MRI with modified-moth-flame algorithm and Kapur's thresholding: a study. Evolutionary Intelligence 2021; 14(2):1053−63.

[24] Fu X, Chen C, Li D. Survival prediction of patients suffering from glioblastoma based on two-branch DenseNet using multi-channel features. International Journal of Computer Assisted Radiology and Surgery 2021; 16(2):207−17.

[25] Mulvey ME. Classification of glioblastoma multiforme genomic subtypes using three-dimensional multiparametric MR imaging features. Doctoral dissertation. San Diego State University; 2016.

[26] Dobson R, Giovannoni G. Multiple sclerosis−a review. European Journal of Neurology 2019;26(1):27−40.

[27] Bakshi R, Thompson AJ, Rocca MA, Pelletier D, Dousset V, Barkhof F, et al. MRI in multiple sclerosis: current status and future prospects. The Lancet Neurology 2008;7(7):615−25.

[28] https://openneuro.org/datasets/ds000115/versions/00001.

[29] Repovs G, Barch DM. Working memory related brain network connectivity in individuals with schizophrenia and their siblings. Frontiers in Human Neuroscience 2012;6:137.

[30] https://wiki.cancerimagingarchive.net/display/Public/RIDER+Collections.

[31] Kadry S, Damaševičius R, Taniar D, Rajinikanth V, Lawal IA. Extraction of tumour in breast MRI using joint thresholding and segmentation−A study. In: 2021 seventh international conference on bio signals, images, and instrumentation (ICBSII). IEEE; March 2021. p. 1−5.

[32] Elanthirayan R, Kubra KS, Rajinikanth V, Raja NSM, Satapathy SC. Extraction of cancer section from 2D breast MRI slice using brain strom

optimization. In: Intelligent data engineering and analytics. Singapore: Springer; 2021. p. 731−9.

[33] Case courtesy of Dr Enrico Citarella, Radiopaedia.org, rID: 39249.

[34] Yu L, Cheng JZ, Dou Q, Yang X, Chen H, Qin J, et al. Automatic 3D cardiovascular MR segmentation with densely-connected volumetric convnets. In: International conference on medical image computing and computer-assisted intervention. Cham: Springer; September 2017. p. 287−95.

[35] Wolterink JM, Leiner T, Viergever MA, Išgum I. Dilated convolutional neural networks for cardiovascular MR segmentation in congenital heart disease. In: Reconstruction, segmentation, and analysis of medical images. Cham: Springer; 2016. p. 95−102.

[36] Lin H, Rajinikanth V. Development of softcomputing tool to evaluate heart MRI slices. International Journal of Computer Theory and Engineering 2019;11(5):80−3.

[37] Manic KS, Hasoon FN, Al Shibli N, Satapathy SC, Rajinikanth V. An approach to examine brain tumor based on Kapur's entropy and Chan−Vese algorithm. In: Third international congress on information and communication technology. Singapore: Springer; 2019. p. 901−9.

[38] Satapathy SC, Rajinikanth V. Jaya algorithm guided procedure to segment tumor from brain MRI. Journal of Optimization 2018;2018.

[39] T Krishnan P, Balasubramanian P, Krishnan C. Segmentation of brain regions by integrating meta heuristic multilevel threshold with markov random field. Current Medical Imaging 2016;12(1):4−12.

[40] Rajinikanth V, Joseph Raj AN, Thanaraj KP, Naik GR. A customized VGG19 network with concatenation of deep and handcrafted features for brain tumor detection. Applied Sciences 2020;10(10):3429.

[41] Rajinikanth V, Priya E, Lin H, Lin F. Hybrid image processing methods for medical image examination. CRC Press; 2021.

[42] Yue Z, Zhao Q, Zhang L, Meng D. Dual adversarial network: toward real-world noise removal and noise generation. In: European conference on computer vision. Cham: Springer; August 2020. p. 41−58.

[43] Agrawal U, Brown EN, Lewis LD. Model-based physiological noise removal in fast fMRI. Neuroimage 2020;205:116231.

[44] Reza AM. Realization of the contrast limited adaptive histogram equalization (CLAHE) for real-time image enhancement. Journal of VLSI Signal Processing Systems for Signal, Image and Video Technology 2004; 38(1):35−44.

[45] Bradley D, Roth G. Adaptive thresholding using the integral image. Journal of Graphics Tools 2007;12(2):13−21.

[46] Bao P, Zhang L, Wu X. Canny edge detection enhancement by scale multiplication. IEEE Transactions on Pattern Analysis and Machine Intelligence 2005;27(9):1485−90.

[47] Marr D, Hildreth E. Theory of edge detection. Proceedings of the Royal Society of London. Series B. Biological Sciences 1980;207(1167):187−217.

[48] Ziou D, Tabbone S. Edge detection techniques-an overview. Pattern Recognition and Image Analysis C/C of Raspoznavaniye Obrazov I Analiz Izobrazhenii 1998;8:537−59.

[49] Deng G, Cahill LW. An adaptive Gaussian filter for noise reduction and edge detection. In: 1993 IEEE conference record nuclear science symposium and medical imaging conference. IEEE; October 1993. p. 1615−9.

[50] Neycenssac F. Contrast enhancement using the Laplacian-of-a-Gaussian filter. CVGIP: Graphical Models and Image Processing 1993;55(6):447−63.

[51] Rajinikanth V, Sivakumar R, Hemanth DJ, Kadry S, Mohanty JR, Arunmozhi S, et al. Automated classification of retinal images into AMD/ non-AMD Class—a study using multi-threshold and Gassian-filter enhanced images. Evolutionary Intelligence 2021;14(2):1163—71.

[52] Guo Z, Zhang L, Zhang D. A completed modeling of local binary pattern operator for texture classification. IEEE Transactions on Image Processing 2010;19(6):1657—63.

[53] Gudigar A, Raghavendra U, Devasia T, Nayak K, Danish SM, Kamath G, et al. Global weighted LBP based entropy features for the assessment of pulmonary hypertension. Pattern Recognition Letters 2019;125:35—41.

[54] Rajinikanth V, Kadry S. Development of a framework for preserving the disease-evidence-information to support efficient disease diagnosis. International Journal of Data Warehousing and Mining (IJDWM) 2021;17(2): 63—84.

[55] Li W, Jia L, Du J. Multi-modal sensor medical image fusion based on multiple salient features with guided image filter. IEEE Access 2019;7: 173019—33.

[56] Khan MA, Kadry S, Alhaisoni M, Nam Y, Zhang Y, Rajinikanth V, et al. Computer-aided gastrointestinal diseases analysis from wireless capsule endoscopy: a framework of best features selection. IEEE Access 2020;8: 132850—9.

[57] Zhang F, Breger A, Cho KIK, Ning L, Westin CF, O'Donnell LJ, et al. Deep learning based segmentation of brain tissue from diffusion MRI. Neuroimage 2021;233:117934.

[58] Akkus Z, Galimzianova A, Hoogi A, Rubin DL, Erickson BJ. Deep learning for brain MRI segmentation: state of the art and future directions. Journal of Digital Imaging 2017;30(4):449—59.

[59] Isın A, Direkoğlu C, Sah M. Review of MRI-based brain tumor image segmentation using deep learning methods. Procedia Computer Science 2016;102:317—24.

4

A study of the segmentation of tumor in breast MRI using entropy thresholding and the Mayfly algorithm

4.1 Introduction

A report of the World Health Organization (WHO) confirmed that the occurrence rate of breast cancer (BC) is gradually increasing [1,2], and preventive methods and early detection can reduce the impact of this disease. In the literature, a considerable amount of research has been published on various cancers and the diagnostic procedures [3–6]. These works also substantiate that all human cancers are controllable when detected at an early stage. These research outcomes also suggest the methodologies regarding various screening procedures to be followed to detect breast disease, such as mammography, ultrasound, thermal imaging, and magnetic resonance imaging (MRI). Compared to other imaging modalities, MRI is one of the commonly used procedures for breast abnormalities due to its flexibility and reputation.

After experiencing a symptom or identifying an abnormal lump with a personal check, the patient will visit the doctor for further assessment and confirmation. The early phase of BC can be detected in hospitals using the suggested medical methods. The doctor initially performs a physical verification of the suspicious area and recommends an image-based assessment. The early stage of BC is due to ductal carcinoma in situ (DCIS) or lobular carcinoma in situ (LCIS) and it can be detected and confirmed using MRI. After detecting the tumor using an image-assisted method, the confirmation of the cancer and its severity will be performed using core-needle biopsy; an invasive technique used to collect and analyze the cancerous tissues to confirm the severity and stage [7,8].

Magnetic Resonance Imaging: Recording, Reconstruction and Assessment. https://doi.org/10.1016/B978-0-12-823401-3.00007-9

Bioimage-supported BC diagnosis has been extensively discussed by scientists and, in each method, semi/automated segmentation is applied to extract the breast tumor (BT) area in the selected image scheme. The appraisal of breast MRI is straightforward and the visibility of tumor fragments is good compared to other breast sections. An evaluation of MRI-based BT can be found in Ref. [9].

In this chapter, the MRI slices collected from Reference Image Database to Evaluate Therapy Response (RIDER) [10] are used for the appraisal. In this work, 12 patient's MRI slices (12 patient $\times$ 15 slices = 180 slices) were extracted using ITK-Snap, which helps to convert 3D MRI into 2D slices. The obtained 2D axial, coronal, and sagittal MRI views were independently inspected using this method and the results are presented and discussed.

Mining of the BT from the chosen 2D MRI is done with the combination of thresholding and segmentation. First, the thresholding course is executed with the Mayfly algorithm and Kapur's entropy (MA + KE) utility and the improved tumor division is then extracted with the watershed (WS) algorithm. This method is an automatic image-mining process that facilitates the extraction of the BT area with improved precision in different views of the MRI slices. After mining the BT, an assessment is applied with respect to the ground truth (GT) and the necessary quality measures (QM) are calculated.

The remainder of this chapter is arranged as follows: Section 4.2 discusses earlier research, Section 4.3 discusses the methodology, and Sections 4.4 and 4.5 give the results and conclusion, respectively.

4.2 Related research

BT is a medical emergency and hence a substantial amount of research has been proposed to observe breast cancer using image-appraisal methods. The previous research on breast irregularities can be classified according to the imaging techniques. The assessment of DCIS with the imaging practice can be accessed from previous works [11−13]. This research employed a selected thresholding and segmentation procedure to mine and assesses the irregular area in MRI.

The previous works on mammography-based abnormality detection can be seen in Refs. [14−16]. In this technique, abnormalities found in breast X-ray images are inspected by means of an appropriate imaging method and the outcomes are demonstrated. Ultrasound is also a noninvasive technique that is employed for breast diagnoses [17]. Breast thermal imaging (BTI)-based cancer discovery is described by Raja et al. [18] and Steven et al. [8]. These research works verify that the early recognition of BC is possible with carefully selected BTI. The segmentation of abnormal breast parts based on the thresholding and segmentation method is reported briefly by Raja et al. [19]. This study used a magnetic resonance angiogram (MRA) image for the appraisal and the obtained results substantiated that the proposed plan helped to achieve an improved result.

In this investigation, a similar method is considered to process MRI slices. To achieve improved results, greater care is taken to choose the thresholding procedure and, with the obtained outcome, the KE-supported enhancement is accepted. Furthermore, to attain improved thresholding, the best threshold process is optimized with the MA. The BT from the preprocessed image is automatically mined with the WS scheme. The required test images were obtained from RIDER-TCIA data and each image was independently inspected through the proposed scheme and the achieved QM confirms the performance of proposed scheme.

4.3 Methodology

Evaluation of breast MRI is essential to recognize illness in patients. The requirement of a suitable image examination methodology is important to mine the infected area from the MRI with improved accuracy.

The employed breast MRI assessment procedure is presented in Fig. 4.1. The different phases of this system include: gathering of 3D scientific grade MRI, 3D to 2D exchange with ITK-Snap, execution of trilevel thresholding with MA + KE to enhance the chosen section, mining of BT with WS, evaluating the BT with GT, and calculating the QM. According to the achieved average QM values, the success of the planned breast MRI tumor assessment method is confirmed.

4.3.1 TCIA-RIDER data set

The performance of an automated disease detection scheme is confirmed using clinical or benchmark images. Most medical images are restricted by ethical/copyright law and approval from the volunteers (patients) is strictly required.

For this reason, clinically collected images are extremely limited for research, and hence most research works are performed on the benchmark data set. In this research, breast

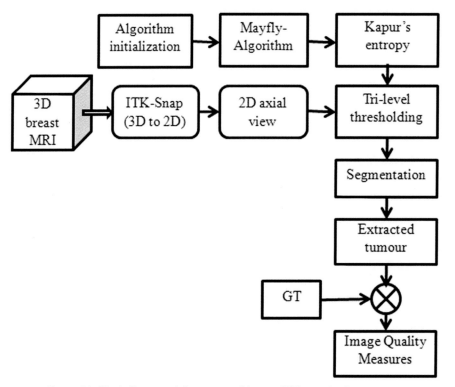

Figure 4.1 Block diagram of the proposed breast MRI examination system.

MRIs available in TCIA-RIDER are adopted and examined. This work considers only 12 patient's images, where from each patient, 15 slices were extracted with ITK-Snap for the investigation. Fig. 4.2 depicts images of three patients with various views considered for the study. The images are in RGB form, and in this work, the grayscale version of images of size $256 \times 256 \times 1$ pixels is used.

Fig. 4.3 shows the test images of 2D MRI slices along with the GT. A related procedure is employed for all other images used in this study.

4.3.2 Trilevel thresholding

Thresholding is a commonly adopted image preprocessing method to enhance the essential areas of an image. Earlier works employed for a number of grayscale/RGB images can be found in Refs. [20,21]. This procedure is applied using a chosen threshold (Th) and, for medical images, trilevel thresholding ($Th = 3$) is employed to separate the given image into the background, normal image area, and tumor; the essential information

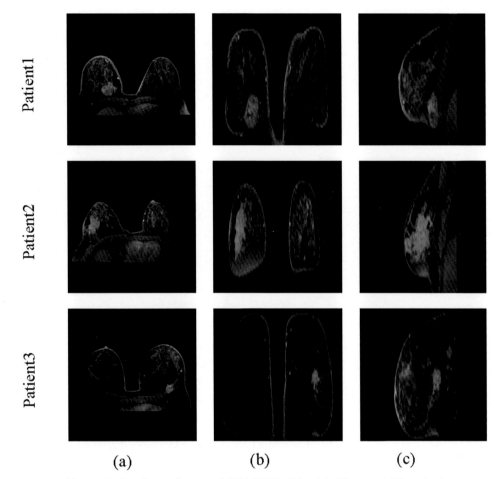

Figure 4.2 Sample test images of TCIA-RIDER: (A) axial; (B) coronal; (C) sagittal.

regarding this process can be found in Refs. [22,23]. Medical images are normally treated using the entropy functions and, in this research work, Kapur's entropy is used to improve the breast MRI. Identification of the KE is achieved using the MA, and other related information about this process can be found in Refs. [24,25].

4.3.2.1 Kapur's entropy

KE has been largely proposed for thresholding grayscale images by means of histogram entropy. This method reveals the finest threshold by maximizing the entropy.

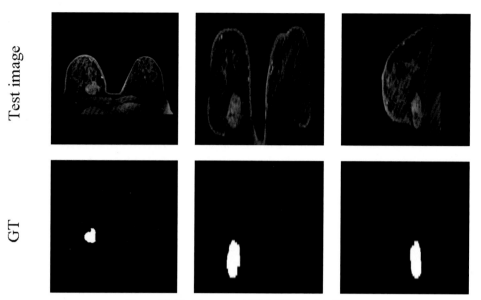

Figure 4.3 Test images of TCIA-RIDER with GT.

For a chosen image, the threshold vector $T = (t_0, t_1, ..., t_{L-1})$, KE is given by: let a chosen grayscale image have L gray-levels and Z total pixels. If $f(k)$ denotes the frequency of the k-th intensity level, then the pixel distribution of the image will be

$$Z = f(0) + f(1) + ... + f(L-1) \tag{4.1}$$

If the probability of the k-th intensity value is given by

$$p(k) = f(k)/Z \tag{4.2}$$

During threshold optimization, the pixels in the image are divided into $Th + 1$ clusters according to the allocated thresholds. After separating the image based on the threshold, the entropy in each bunch is individually premeditated and united to get the ultimate entropy as follows:

$$\text{Bi} - \text{level threshold} = f(t_1, t_2) = e_0 + e_1 \tag{4.3}$$

$$\text{Multi} - \text{level threshold} = f(t_1, t_2, ...t_L) = e_0 + e_1 + ... + e_{L-1} \tag{4.4}$$

$$e_0 = - \sum_{k=0}^{k=t_{1-1}} \frac{p_k}{\sigma_0} \ln \frac{p_k}{\sigma_0}, \sigma_0 = \sum_{k=0}^{k=t_{1-1}} p_k$$

$$e_1 = -\sum_{k=t_{1-1}}^{k=t_{1-2}} \frac{p_k}{\sigma_1} \ln \frac{p_k}{\sigma_1}, \sigma_1 = \sum_{k=t_{1-1}}^{k=t_{1-2}} p_k \qquad (4.5)$$

$$e_{L-1} = -\sum_{k=t_{L-1}}^{k=t_{L-2}} \frac{p_k}{\sigma_{L-1}} \ln \frac{p_k}{\sigma_{L-1}}, \sigma_{L-1} = \sum_{k=t_{L-1}}^{k=t_{L-2}} p_k$$

where e = entropy, p = probability distribution, and σ = probability occurrence.

$$\mathrm{Kapur}_{\max}(T) = \sum_{p=1}^{L-1} H_j^C \qquad (4.6)$$

Other information on KE can be found in Refs. [26,27].

4.3.2.2 Mayfly algorithm

MA is a newly proposed nature-motivated method developed by integrating the merits of other schemes, such as particle swarm optimization (PSO), firefly algorithm (FA), and genetic algorithm (GA) [28]. The various stages of MA include: (1) initialization of male and female flies, (2) authorize the male Mayfly to notice the G_{best} for the thresholding problem, (3) allow the female Mayfly to recognize and unite with the male Mayfly awaiting in G_{best}, (4) offspring formation, and (5) stopping the investigation and delivering the best outcome [29–31].

The exploration capability of MA relies on the premature position of the male and its distance from the female. When each fly is arbitrarily placed in a selected investigative site with a suitable number of male and female agents, all flies are allowed to attain G_{best} according to the convergence (Fig. 4.4). This procedure continues until an identical quantity of offspring is generated with each pair of agents. To finish the exploration of the MA at this phase, the selection preference for the offspring is allocated with zero velocity. Essential details of the MA can be found in Refs. [28–31].

Fig. 4.5 shows the initial situation of the male Mayfly in a selected investigative position. The number of male (X) and female (Y) flies is allocated with a selected number (N). Fig. 4.5 presents the union of a male fly toward the best position and Fig. 4.6 depicts the final outcome of the MA. To demonstrate the working of MA, let there be an equal number of male (M) and female (F) flies arbitrarily initialized in a D-dimensional investigative space and each fly is depicted as follows: $i = 1, 2, ..., N$

Initialization

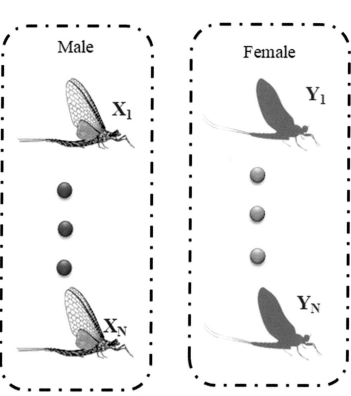

Figure 4.4 Initial positions of mayflies in a chosen search space.

(for $N = 25$). During this investigation, each agent is arbitrarily located in this search space and each fly is allowed to go toward the best location (G_{best}). As in Fig. 4.5 the male fly is allowed to converge at G_{best} by changing its position and velocity. The convergence of the fly toward the best outcome is monitored by the Cartesian distance along with an increase in iteration. This process is similar to the movement of the fly in FA. The position and velocity update expressions are shown in Eqs. (4.7) and (4.8):

$$E_i^{t+1} = E_i^t + F_i^{t+1} \qquad (4.7)$$

$$F_{i,j}^{t+1} = F_{i,j}^t + C_1 {}^* \bar{e}^{-\beta D_p 2} \left(P_{best_{i,j}} - E_{i,j}^t \right) + C_2 {}^* \bar{e}^{-\beta D_g 2} \left(G_{best_{i,j}} - E_{i,j}^t \right) \qquad (4.8)$$

where E_i^t and E_i^{t+1} are the primary and final positions, F_i^{t+1} and $F_{i,j}^{t+1}$ are the primary and end velocities, respectively, local learning limitation $(C_1) = 1$, global learning limitation

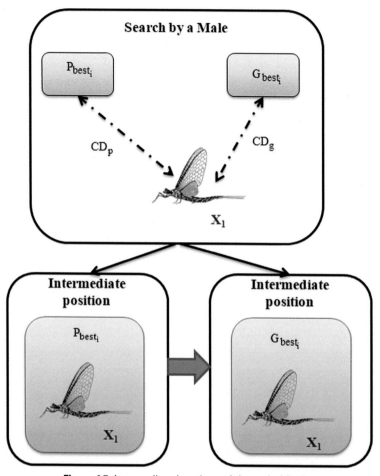

Figure 4.5 Intermediate locations of the male Mayfly.

$(C_2) = 1.5$, $\beta = 2$ and D_p and D_g are the Cartesian distance. Eq. (4.2) is created by combining FA and PSO. When the updating of the flies continues, every M will achieve G_{best} and execute a velocity updating to attract the F by performing an inimitable nuptial dance-like movement.

The velocity revision throughout this procedure can be defined as in Eq. (4.9);

$$F_{i,j}^{t+1} = F_{i,j}^{t} + d^{*}R \qquad (4.9)$$

where the nuptial dance value $(d) = 5$ and $R =$ a random number $[-1,1]$.

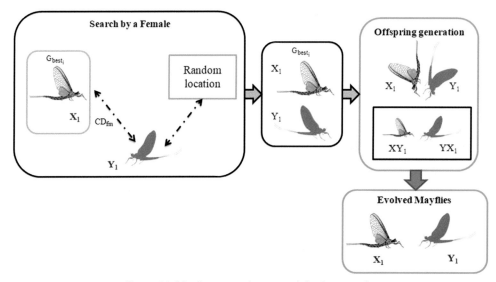

Figure 4.6 Mayfly converging toward the best result.

When the search of M is completed, every F is then permitted to find an M converged atG_{best}. Fig. 4.6 shows the investigation by F, which travels toward M based on a distance $\left(D_{mf}\right)$ or the Y escapes to a new section with a random walk $(W = 1)$.

The position and velocity updates in F are depicted in Eqs. (4.10) and (4.11).

$$E_i'^{t+1} = E_i'^t + F_i'^{t+1} \tag{4.10}$$

$$F_{i,j}'^{t+1} = \begin{cases} F_{i,j}'^t + C_2 e^{-\beta D_{mf}^2}\left(M_{i,j}^t - Y_{i,j}^t\right) & \text{if } O(F_i) > O(M_i) \\ F_{i,j}'^t + W^*r & \text{if } O(F_i) \leq O(M_i) \end{cases} \tag{4.11}$$

where O = objective function.

When the iteration increases, every F will reach M to generate the offspring as depicted in Fig. 4.6. After this generation, every offspring is discarded using a weight zero and this process continues until the MA is terminated. In this research, the MA is employed to find the best threshold in the MRI slice by maximizing the KE.

4.3.2.3 Thresholding

Thresholding is a proven image enhancement procedure in which a chosen methodology is employed to enhance the image

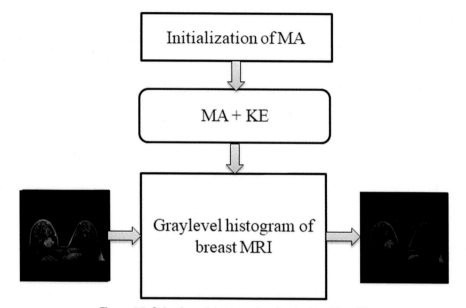

Figure 4.7 Selection of the best threshold with MA + KE.

by grouping the pixels based on need. The procedure employed in this work is shown in Fig. 4.7 and the achieved answer for $Th = 3$ shows the enhanced visibility of the tumor compared with the test image. This process is applied to other images considered in this research and the outcome is then treated using the segmentation scheme, which extracts the tumor with better visibility.

4.3.3 Segmentation

Mining the necessary section from the raw/preprocessed image is necessary to detect the disease and its severity using breast MRI. A number of semi/automated segmentation procedures are described in the literature, and in this work, watershed (WS), level set (LS) and active contour (AC) methods are considered to extract the tumor from the MRI.

The watershed is an automated approach consisting of procedures, such as edge detection, watershed fill, morphology-assisted enhancement, localization, and mining. In MS, the morphological operation plays a vital role in extracting the tumor, and the earlier works on WS can be found in Refs. [32–34].

The LS and AC methods are semiautomated procedures based on a bounding box. When the operator initiates the box on a chosen section, this box adjusts its positions until it identifies all the necessary pixels of the image to be mined. This adjustment of the box is performed based with an increase in the iteration and this process stops only when all the pixels are identified completely. After the convergence, the extracted segment will be in a binary form (combination of black and white pixels), and then it is compared with the GT to find the QM. Based on the attained values of the QM, the performance of the proposed breast MRI assessment scheme is confirmed.

4.3.4 Assessment and validation

The medical imaging literature confirms that the performance of a disease detection scheme must be assessed to confirm its clinical significance. In this work, the performances of MA + KE thresholding and WS/LS/AC segmentation are individually evaluated by computing the necessary QM, as depicted in Eqs. (4.12) −(4.17).

The pixel-level comparison of segmented tumor and GT helps to compute the true-positive (*TP*), false-negative (*FN*), true-negative (*TN*), false-positive (*FP*), positive-pixel (*P = TP + FN*), and negative-pixel (*N = TN + FP*) values. From these values, other QM, such as Jaccard (JA), dice (DI), accuracy (AC), precision (PR), sensitivity (SE), and specificity (SP) are computed and, based on the obtained values, the superiority of the breast MRI detection is confirmed.

The mathematical representations of these values are [35]:

$$JI = \frac{TP}{TP + FP + FN} \tag{4.12}$$

$$DC = \frac{2TP}{2TP + FN + FP} \tag{4.13}$$

$$AC = \frac{TP + TN}{TP + TN + FP + FN} \tag{4.14}$$

$$PR = \frac{TP}{TP + FP} \tag{4.15}$$

$$SE = \frac{TP}{TP + FN} \tag{4.16}$$

$$SP = \frac{TN}{TN + FP} \tag{4.17}$$

4.4 Result and discussion

This section describes the experimental results obtained using the breast MRI test images of TCIA-RIDER. These results were obtained using MATLAB.

Initially, the coronal-view MRI is considered for the assessment, and the schemes such as MA + KE-based enhancement and WS/LS/AC-based segments are applied. The outcomes attained in each case are separately recorded and evaluated. Initially, the proposed work implements the MA and KE-based tri-level thresholding process on the chosen image and, during this search, the following values are assigned for the MA: number of flies = 25, search dimension = 3, maximum iteration = 5000, stopping criteria = maximal iteration or maximal KE. When this process is implemented, after finding the maximized KE, this search stops and presents the images, such as preprocessed MRI, optimal threshold, and maximized KE.

Fig. 4.8 presents the outcome of the entropy with respect to the iteration value. This presents the maximal value of the KE attained

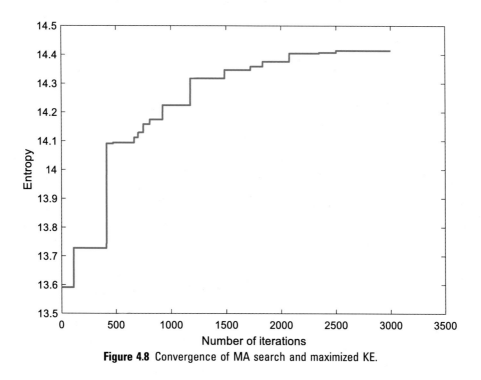

Figure 4.8 Convergence of MA search and maximized KE.

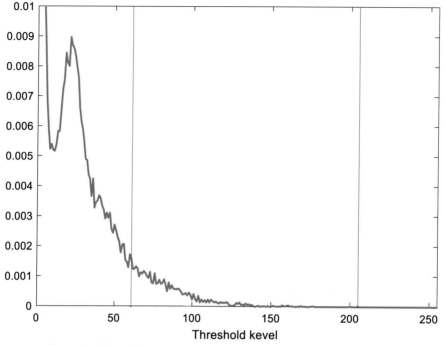

Figure 4.9 Trilevel threshold separation of the histogram using MA + KE.

for the image for $Th = 3$. The corresponding threshold value is depicted in Fig. 4.9, and this shows the grouped threshold for $Th = 3$. This grouping helps to obtain an image such as the tumor, normal area, and background. After enhancing the image, the tumor section is extracted using the WS/LS/AC procedures. Fig. 4.10 depicts the outcome of the segmentation process, in which Fig. 4.10A–D show the outcomes, such as detected edge, watershed fill, morphological operation, and extracted tumor, respectively. Fig. 4.10E–H present the outcome attained with the LS and AC, such as converged LS, extracted tumor, converged AC, and extracted tumor, respectively.

Fig. 4.11 presents the attained results for WS, LS, and AC for all the considered MRI views (axial, coronal, and sagittal) and then every binary image of the tumor is compared with its GT, and the QM are computed. The attained results for the sample test images with every chosen procedure are illustrated in Tables 4.1 and 4.2.

Table 4.1 presents the pixel-level information and Table 4.2 shows the QM attained with the values of Table 4.1. This information confirms that the proposed method helps to get better values of JI, DC, and AC for the chosen sample test image.

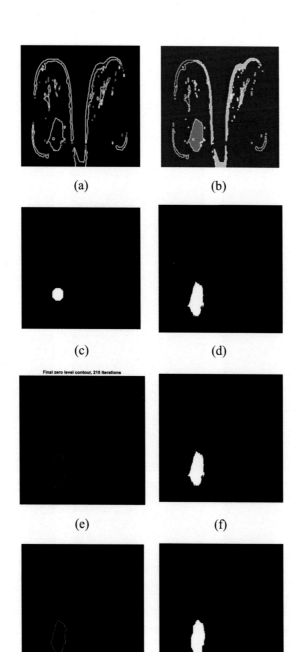

Final zero level contour, 210 iterations

(a)   (b)

(c)   (d)

(e)   (f)

(g)   (h)

Figure 4.10 Outcomes attained with the chosen segmentation process: (A) edge detection; (B) watershed; (C) initial morphology; (D) tumor by WS; (E) LS convergence; (F) tumor by LS; (G) AC convergence; (H) tumor by AC.

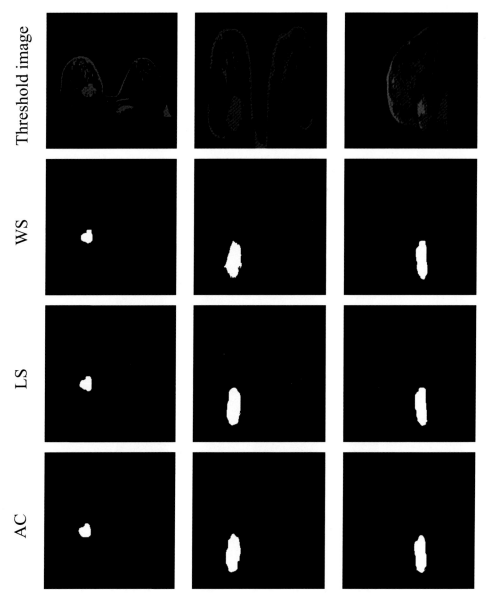

Figure 4.11 Segmented tumor sections from breast MRI.

Figs. 4.12–4.14 present the Glyph-plot-based comparisons of the outcomes for various MRI views and these results confirm that the overall outcome attained with the LS technique is better than with WS and AC.

Fig. 4.15 presents the average values of the performance measures computed for all the patients' MRI slices (180 in total) and

Table 4.1 Pixel-level quality measures computed during tumor and GT comparison.

Method	MRI view	Pixels					
		TP	FN	TN	FP	P	N
WS	Axial	423	16	64,990	107	439	65,097
	Coronal	1423	43	63,825	245	1466	64,070
	Sagittal	1278	11	64,071	176	1289	64,247
LS	Axial	471	21	64,985	59	492	65,044
	Coronal	1492	40	63,828	176	1532	64,004
	Sagittal	1301	28	64,054	153	1329	64,207
AC	Axial	450	18	64,988	80	468	65,068
	Coronal	1490	65	63,803	178	1555	63,981
	Sagittal	1281	26	64,056	173	1307	64,229

Table 4.2 Quality measures derived from the pixel-level information during tumor and GT comparison.

Method	View	JI	DC	AC	PR	SE	SP
WS	Axial	77.4725	87.3065	99.8123	79.8113	96.3554	99.8356
	Coronal	83.1677	90.8105	99.5605	85.3118	97.0668	99.6176
	Sagittal	87.2355	93.1826	99.7147	87.8955	99.1466	99.7261
LS	Axial	85.4809	92.1722	99.8779	88.8679	95.7317	99.9093
	Coronal	87.3536	93.2500	99.6704	89.4484	97.3890	99.7250
	Sagittal	87.7868	93.4962	99.7238	99.7238	97.8932	99.7617
AC	Axial	82.1168	90.1804	99.8505	84.9057	96.1538	99.8771
	Coronal	85.9781	92.4604	99.6292	89.3285	95.8199	99.7218
	Sagittal	86.5541	92.7925	99.6964	88.1018	98.0107	99.7307

WS   LS   AC

Figure 4.12 Glyph-plot of the performance measures attained with an axial-view MRI slice.

WS   LS   AC

Figure 4.13 Glyph-plot of the performance measures attained with a coronal-view MRI slice.

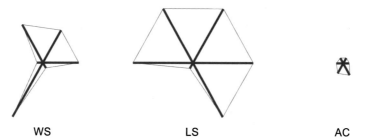

WS   LS   AC

Figure 4.14 Glyph-plot of the performance measures attained with a sagittal-view MRI slice.

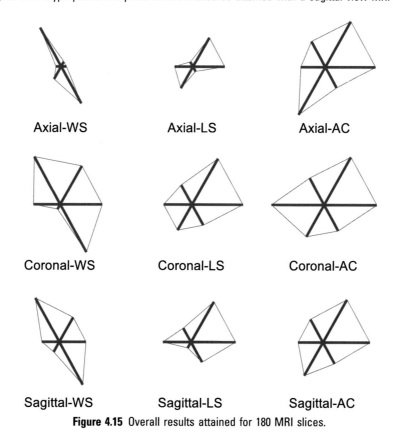

Axial-WS   Axial-LS   Axial-AC

Coronal-WS   Coronal-LS   Coronal-AC

Sagittal-WS   Sagittal-LS   Sagittal-AC

Figure 4.15 Overall results attained for 180 MRI slices.

this result confirms that the proposed thresholding (MA + KE) and segmentation (WS/LS/AC) helps to obtain better quality measures. The average results obtained with the proposed scheme on all the images are very similar, and this confirms that the proposed work is clinically significant and suitable for the extraction of tumor in axial, coronal and sagittal views of MRI slices.

4.5 Conclusion

Early and precise recognition of the breast cancer area will support suitable treatment. This work has employed a combined preprocessing (thresholding) and segmentation technique to assess breast MRI slices of the TCIA-RIDER data set. Herein, 180 MRI slices with axial, coronal, and sagittal views were examined using the proposed method. This work implemented MA + KE-based trilevel thresholding and WS/LS/AC segmentation to obtain the tumor. After mining the BT from the MRI, an assessment with GT was executed and the necessary QM computed. The overall outcome achieved through the proposed research on the different views of MRI slices confirms that this system is clinically noteworthy and that in the future it could be used to assess clinically collected breast MRIs.

References

[1] DeSantis CE, Bray F, Ferlay J, Lortet-Tieulent J, Anderson BO, Jemal A. International variation in female breast cancer incidence and mortality rates. Cancer Epidemiology and Prevention Biomarkers 2015;24(10): 1495−506.

[2] https://www.who.int/news-room/fact-sheets/detail/breast-cancer.

[3] Rajinikanth V, Kadry S, Taniar D, Damaševičius R, Rauf HT. Breast-cancer detection using thermal images with marine-predators-algorithm selected features. In: 2021 seventh international conference on bio signals, images, and instrumentation (ICBSII). IEEE; March 2021. p. 1−6.

[4] Kadry S, Damaševičius R, Taniar D, Rajinikanth V, Lawal IA. Extraction of tumour in breast MRI using joint thresholding and segmentation−A study. In: 2021 seventh international conference on bio signals, images, and instrumentation (ICBSII). IEEE; March 2021. p. 1−5.

[5] Zemmal N, Azizi N, Dey N, Sellami M. Adaptive semi supervised support vector machine semi supervised learning with features cooperation for breast cancer classification. Journal of Medical Imaging and Health Informatics 2016;6(1):53−62.

[6] Bhattacherjee A, Roy S, Paul S, Roy P, Kausar N, Dey N. Classification approach for breast cancer detection using back propagation neural network: a study. In: Deep learning and neural networks: concepts, methodologies, tools, and applications. IGI Global; 2020. p. 1410−21.

[7] Virmani J, Dey N, Kumar V. PCA-PNN and PCA-SVM based CAD systems for breast density classification. In: Applications of intelligent optimization in biology and medicine. Cham: Springer; 2016. p. 159−80.

[8] Fernandes SL, Rajinikanth V, Kadry S. A hybrid framework to evaluate breast abnormality using infrared thermal images. IEEE Consumer Electronics Magazine 2019;8(5):31−6.

[9] Elanthirayan R, Kubra KS, Rajinikanth V, Raja NSM, Satapathy SC. Extraction of cancer section from 2D breast MRI slice using brain strom optimization. In: Intelligent data engineering and analytics. Singapore: Springer; 2021. p. 731−9.

[10] https://wiki.cancerimagingarchive.net/display/Public/RIDER+Collections.

[11] Rajinikanth V, Raja NSM, Satapathy SC, Dey N, Devadhas GG. Thermogram assisted detection and analysis of ductal carcinoma in situ (DCIS). In: 2017 international conference on intelligent computing, instrumentation and control technologies (ICICICT). IEEE; July 2017. p. 1641−6.

[12] Holland R, Peterse JL, Millis RR, Eusebi V, Faverly D, van de Vijver MA, Zafrani B. Ductal carcinoma in situ: a proposal for a new classification. Seminars in Diagnostic Pathology August 1994;11(3):167−80.

[13] Groen EJ, Elshof LE, Visser LL, Emiel JT, Winter-Warnars HA, Lips EH, Wesseling J. Finding the balance between over-and under-treatment of ductal carcinoma in situ (DCIS). The Breast 2017;31:274−83.

[14] Salem MAM. Mammogram-Based cancer detection using deep convolutional neural networks. In: 2018 13th international conference on computer engineering and systems (ICCES). IEEE; December 2018. p. 694−9.

[15] Singh AK, Gupta B. A novel approach for breast cancer detection and segmentation in a mammogram. Procedia Computer Science 2015;54: 676−82.

[16] Rodríguez-Ruiz A, Krupinski E, Mordang JJ, Schilling K, Heywang-Köbrunner SH, Sechopoulos I, Mann RM. Detection of breast cancer with mammography: effect of an artificial intelligence support system. Radiology 2019;290(2):305−14.

[17] Raj SS, Raja NSM, Madhumitha MR, Rajinikanth V. Examination of digital mammogram using otsu's function and watershed segmentation. In: 2018 fourth international conference on biosignals, images and instrumentation (ICBSII). IEEE; 2018, March. p. 206−12.

[18] Raja N, Rajinikanth V, Fernandes SL, Satapathy SC. Segmentation of breast thermal images using Kapur's entropy and hidden Markov random field. Journal of Medical Imaging and Health Informatics 2017;7(8):1825−9.

[19] Raja NSM, Sukanya SA, Nikita Y. Improved PSO based multi-level thresholding for cancer infected breast thermal images using Otsu. Procedia Computer Science 2015;48:524−9.

[20] Kadry S, Rajinikanth V, Raja NSM, Hemanth DJ, Hannon NM, Raj ANJ. Evaluation of brain tumor using brain MRI with modified-moth-flame algorithm and Kapur's thresholding: a study. Evolutionary Intelligence 2021; 14(2):1053−63.

[21] Rajinikanth V, Raja NSM, Dey N. A beginner's guide to multilevel image thresholding. CRC Press; 2020.

[22] Pugalenthi R, Rajakumar MP, Ramya J, Rajinikanth V. Evaluation and classification of the brain tumor MRI using machine learning technique. Journal of Control Engineering and Applied Informatics 2019;21(4):12−21.

[23] Hore S, Chakraborty S, Chatterjee S, Dey N, Ashour AS, Van Chung L, Le DN. An integrated interactive technique for image segmentation using stack based seeded region growing and thresholding. International Journal of Electrical and Computer Engineering 2016;6(6):2088−8708.

[24] Kapur JN, Sahoo PK, Wong AK. A new method for gray-level picture thresholding using the entropy of the histogram. Computer Vision, Graphics, and Image Processing 1985;29(3):273−85.

[25] Dey N, Rajinikanth V, Hassanien AE. An examination system to classify the breast thermal images into early/acute DCIS class. In: Proceedings of

international conference on data science and applications. Singapore: Springer; 2021. p. 209—20.

[26] Dey N, Rajinikanth V, Fong SJ, Kaiser MS, Mahmud M. Social group optimization—assisted Kapur's entropy and morphological segmentation for automated detection of COVID-19 infection from computed tomography images. Cognitive Computation 2020;12(5):1011—23.

[27] Rajinikanth V, Dey N, Kavitha S. Multi-thresholding with Kapur's entropy—a study using bat algorithm with different search operators. In: Applications of bat algorithm and its variants. Singapore: Springer; 2021. p. 61—78.

[28] Zervoudakis K, Tsafarakis S. A mayfly optimization algorithm. Computers and Industrial Engineering 2020;145:106559.

[29] Bhattacharyya T, Chatterjee B, Singh PK, Yoon JH, Geem ZW, Sarkar R. Mayfly in harmony: a new hybrid meta-heuristic feature selection algorithm. IEEE Access 2020;8:195929—45.

[30] Gao ZM, Zhao J, Li SR, Hu YR. The improved mayfly optimization algorithm. Journal of Physics: Conference Series November 2020;1684(1): 012077.

[31] Zhao J, Gao ZM. The negative mayfly optimization algorithm. Journal of Physics: Conference Series, 1693. IOP Publishing; December 2020. p. 012098. 1.

[32] Huang YL, Chen DR. Watershed segmentation for breast tumor in 2-D sonography. Ultrasound in Medicine and Biology 2004;30(5):625—32.

[33] Hamarneh G, Li X. Watershed segmentation using prior shape and appearance knowledge. Image and Vision Computing 2009;27(1—2):59—68.

[34] Arunmozhi S, Raja NSM, Rajinikanth V, Aparna K, Vallinayagam V. Schizophrenia detection using brain MRI—a study with watershed algorithm. In: 2020 international conference on system, computation, automation and networking (ICSCAN). IEEE; July 2020. p. 1—4.

[35] Rajinikanth V, Raja NSM, Kamalanand K. Firefly algorithm assisted segmentation of tumor from brain MRI using Tsallis function and Markov random field. Journal of Control Engineering and Applied Informatics 2017; 19(3):97—106.

5

Abnormality detection in heart MRI with spotted hyena algorithm-supported Kapur/Otsu thresholding and level set segmentation

5.1 Introduction

The heart is one of the most important organs in human physiology and abnormalities of the heart can caused problems in the blood circulatory system, which affects the oxygen supply. These abnormalities can be due to a variety of reasons and can be commonly identified using a biosignal (ECG)-supported diagnosis [1−3]. Along with the biosignal, bioimage-based heart detection is also adopted to detect diseases/abnormalities in the heart.

Magnetic resonance imaging (MRI) is a commonly recommended imaging scheme in hospitals and is extensively adopted in medical clinics to record the performance of internal organs. Usually, MRI recording is carried out by a skilled radiologist, and the final structure of the MRI is accessible as a reconstructed three-dimensional (3D) illustration. The advantages of MRI compared to other bioimaging methods are as follows: (1) obtainable as a 3D image, (2) accessible with diverse modalities, such as flair, T1, T1C, T2, DE, and fMRI, and (3) can be used with or without a contrast agent [4−6].

When a patient is scanned using MRI, the raw image collected from the scanning process is in 3D form and, for simplicity, it can be converted into a 2D slice and the axial view can be considered for computer-based or visual verification by a doctor. Although the information available in 3D images is superior, it needs extra computational effort to inspect the disease details. Hence, in most cases, the 3D image is converted into 2D axial, coronal and sagittal view slices, and verified physically using the available

Magnetic Resonance Imaging: Recording, Reconstruction and Assessment. https://doi.org/10.1016/B978-0-12-823401-3.00006-7

disease score by a radiologist and an examination by a doctor. In the majority of cases, 2D slices of the top view (axial) are most extensively used due to its advantages.

The aim of this research is to develop a segmentation scheme to extract and evaluate the heart MRI (HMRI). The HMRI is used in cardiac centers to show the heart's anatomy, functioning, and cardiac abnormalities. The HMRI is obtained in 3D form and a procedure is used to convert the 3D image into 2D slices. This research uses an examination scheme for abnormal areas in HMRI using a selected imaging practice. The preprocessing is employed to threshold the chosen MRI slice to improve the visibility of the heart area to be assessed. This work implemented a trilevel threshold with the spotted hyena algorithm (SHA) and Kapur/fuzzy/Otsu function. Initially, this process is achieved using Kapur's entropy (KE), and then it is repeated using fuzzy entropy (FE) and Otsu's between-class variance (BCV). The performance of the thresholding is verified by calculating the image quality values (IQV) computed using a comparison between the original and threshold images.

The abnormal area enhanced with the threshold is mined by means of a semiautomated method; level set (LS) segmentation. This procedure is a bounding-box-based approach, and the box is permitted to converge toward the abnormal area pixels when iteration increases and this process stops when all the pixels have been identified by the gradually adjusting contour. The extracted section is in binary form (0 pixels for background and 1 pixels for the section) and then it is compared with the GT, and the necessary quality measures (QM) are computed and, based on the attained value, the performance of the proposed scheme is confirmed.

This chapter uses e test HMRI images collected from HVSMR2016's benchmark images for assessment [7–9]. Herein, 10 volunteers' images are considered and 150 images (10 patient $\times$ 15 slices) with dimensions of $256 \times 256 \times 1$ pixels are considered for the evaluation, with the results separately presented for SHA + KE + LS, SHA + FE + LS, and SHA + BCV + LS and based on the computed QM, the performance of the proposed scheme is confirmed.

The other sections of this chapter are arranged as follows: Section 5.2 describes the related research, Section 5.3 shows the methods considered to develop the system, and the experimental outcome and conclusion are presented in Sections 5.4 and 5.5, respectively.

5.2 Related research work

Recently, a substantial amount of MRI assessment tasks have been carried out by researchers. The majority of these works are associated with MRI-supported brain abnormality detection and they can be adopted to examine the HMRI.

The research by Guttman et al. methods to be implemented during the assessment of HMRI [10]. Backhaus et al. discussed a practice for automated quantification of biventricular volumes and function by means of HMRI [11]. The research of Khan et al. offered a comprehensive appraisal of HMRI assessment procedures presented in the research work of 100 articles [12]. Friedrich demonstrated a detailed procedure to appraise and outline the future scope of HMRI and its clinical applications [13]. The investigation by Attili et al. demonstrated the quantification practice to be followed for HMRI examination during abnormality detection [14]. The research by Gupta et al. presented a thorough assessment of the HMRI during infection classification practice [15]. Along with the precise HMRI assessment procedures, brain MRI assessment practices employed in the literature also adopt the related method to obtain the irregular area from the HMRI of chosen modalities.

In the proposed work, HMRI recorded with the Fl-modality is considered for the assessment and the considered axial-view (2D slice) image is examined using the joint thresholding and segmentation process. This work is a soft computing method and uses the traditional SHA to preprocess the image based on a chosen thresholding scheme. Finally, the abnormal section is extracted with LS and evaluated against GT to confirm the merit of the implemented scheme.

5.3 Methodology

This section of the chapter presents the information regarding the various image-processing methods employed to extract the abnormal section from an HMRI. Fig. 5.1 presents the cardiac abnormality examination scheme using HMRI. Initially, ITK-Snap was used to convert the 3D MRI into a 2D slice and, in this work, flair-weighted modality MRI is used for the examination. The proposed scheme is implemented on an axial-view image and the required section was extracted using the combined thresholding and segmentation procedure. The thresholding was initially implemented to enhance the abnormal area with the help of a trilevel thresholding. The identification of the finest

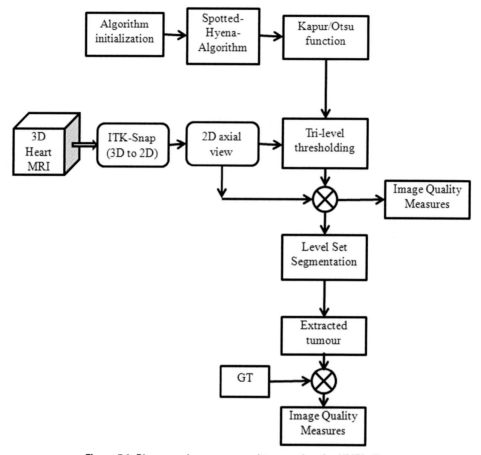

Figure 5.1 Disease scheme proposed to examine the HMRI slice.

threshold was achieved using SHA and Kapur/Otsu. Initially, Kapur's approach was used to improve the image and the improved area was then extracted using the LS segmentation. Later, a similar process was repeated using Otsu and the outcomes of Kapur and Otsu were separately tested and compared. In this work, two level performance comparisons for (1) thresholding outcome and (2) segmentation outcome are separately computed and evaluated using the HMRI images of the HVSMR2016 data set.

5.3.1 MRI database

The performance of the proposed image examination scheme is evaluated and its clinical significance be confirmed using a clinical-grade MRI. In this research, the necessary cardiac MRI was collected from the benchmark grand-challenge database, HVSMR2016, and the 2D slices of this database were used for the evaluation. From HVSMR2016, 10 volunteers' images were used for the assessment and from every image set, 15 slices with dimensions of $174 \times 118 \times 4$ pixels were initially extracted and then converted into $256 \times 256 \times 1$ pixels for the examination. A similar procedure was followed for the GT images available in the database and the developed image examination task was implemented on the HMRI flair modality slices. The sample test images and GT are depicted in Fig. 5.2.

5.3.2 Preprocessing

Image preprocessing is a necessary procedure to be employed to correct or enhance a test image to the required level. Herein, trilevel thresholding based on Kapur/Otsu is implemented to improve the visibility of the abnormal heart section. Identification of the best threshold is achieved using the SHA and the performance of the Kapur/Otsu scheme is compared and validated.

The mathematical expression of Kapur's entropy (KE) was discussed in Chapter 4 (subsection 4.3.2.1) and the maximization of KE is considered as a task during the SHA algorithm's optimal threshold search. The theory for fuzzy entropy (FE) can be accessed from earlier research works [16–19].

5.3.2.1 Otsu's function

The BCV is the Otsu's nonparametric thresholding process, which must be attained by arbitrarily changing thresholds in the image using SHA.

Let us consider the grayscale imagery for the conversation; for bilevel operation, threshold is chosen as: $Th = th_0, th_1$ to split the image into two groups (i.e., G_0 and G_1). The category G_0 includes the pixels of range 0 to th_0 and class G_1 includes th_1 to 255. This discussion can be expressed based on its probability distribution task and its allocation for G_0 and G_1 can be presented as

$$G_0 = \frac{P_0}{\eta_0(Th)} \cdots \frac{P_{th_0-1}}{\eta_0(Th)} \text{ and } G_1 = \frac{P_{th_0}}{\eta_1(Th)} \cdots \frac{p_{255}}{\eta_1(Th)} \quad (5.1)$$

where $\eta_0(Th) = \sum_{i=0}^{Th-1} P_i$, $\eta_1(Th) = \sum_{i=Th}^{255} P_i$

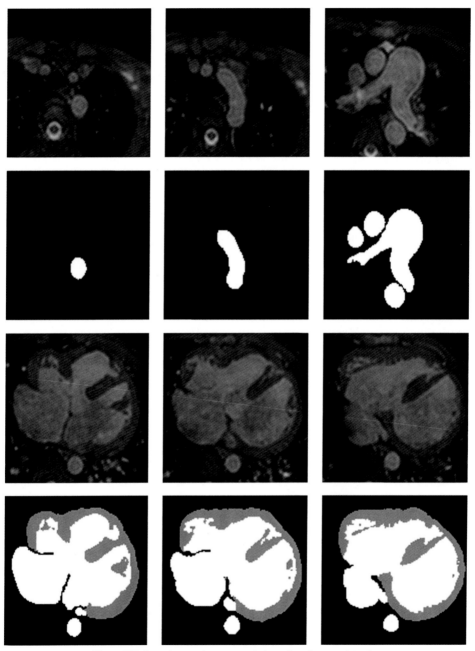

Figure 5.2 Sample test images and the related ground-truth.

The mean values; ψ_0, ψ_1 for G_0, G_1 can be presented as

$$\psi_0 = \sum_{i=0}^{Th-1} \frac{iP_i}{\eta_0(Th)} \text{ and } \psi_1 = \sum_{i=Th}^{255} \frac{iP_i}{\eta1(Th)} \qquad (5.2)$$

The mean intensity (ψ_{Th}) of the whole image can be represented as

$\psi_{Th} = \eta_0\psi_0 + \eta_1\psi_1$ and $\eta_0 + \eta_1 = 1$.

The guiding value for bilevel thresholding is expressed as

$$\text{Otsu}_{max} = J(Th) = \vartheta_0 + \vartheta_1 \qquad (5.3)$$

where $\vartheta_0 = \eta_0(\psi_0 - \psi_{Th})^2$ & $\vartheta_1 = \eta_1(\psi_1 - \psi_{Th})^2$.

Eqs. (5.1) and (5.2) denote the initial parameters and Eq. (5.3) can be changed to a multithreshold problem by changing "Th" values, as in Eq. (5.4):

$$\text{Otsu}_{max} = J(Th) = \vartheta_0 + \vartheta_1 + \cdots + \vartheta_{L-1} \qquad (5.4)$$

where $\vartheta_0 = \eta_0(\psi_0 - \psi_{Th})^2, \vartheta_1 = \eta_1(\psi_1 - \psi_{Th})^2, ..., \vartheta_{Th}$
$= \eta_{Th}(\psi_{Th} - \psi_{L-1})^2$

According to the requirements, the threshold value can be chosen as $Th = 2, 3, 4, ...L - 1$. Other information on Otsu can be found in Reference [20].

5.3.2.2 Spotted hyena algorithm

The SHA is a nature-inspired optimization technique developed in 2017 by mimicking the hunting actions of a spotted hyena pack. The spotted hyena is a skilled hunter and normally lives in a group. The hunting tactic followed by hyenas consists of: (1) selecting and tracking the prey, (2) chasing the prey as a group, (3ii) Encircling, and (4) killing the prey. The mathematical model developed by Dhiman and Kumar [21] considered all the possible constraints to improve the convergence capability of the SHA. This algorithm is developed by balancing the exploration and exploitation operations and helped to achieve the optimal results for a number of engineering problems [22−24].

5.3.2.2.1 Encircling

The leader of the pack will identify the prey by sight, sound, and smell, and, after selecting the prey, the leader and its pack will chase the prey until it is tired. When the prey is tired, the leader and the group will encircle it, as depicted in Fig. 5.3. In

Figure 5.3 Encircling process of the hyena pack toward its prey.

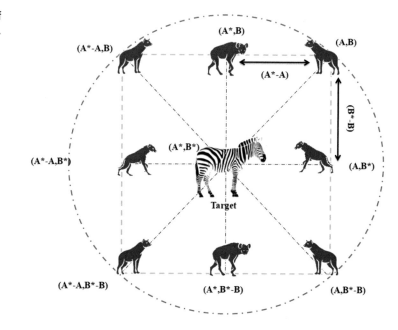

this every member of the group will adjust its position/distance with respect to the prey, and this process is depicted in the figure using the notation A and B, this adjustment is carried out in the algorithm using mathematical operations such as multiplication and subtraction.

The encircling process is mathematically represented in Eqs. (5.5)–(5.8);

$$\vec{D}_h = \left| \vec{B} \cdot \vec{P}_p(x) - \vec{P}(x) \right| \tag{5.5}$$

$$\vec{P}(x+1) = \vec{P}_p(x) - \vec{E} \cdot \vec{D}_h \tag{5.6}$$

where $\vec{D}_h$ = distance between the hyena and prey, x = current iteration, $\vec{P}_p$ = position vector of prey, and $\vec{P}$ = position vector of hyena.

The coefficient vectors $\vec{B}$ and $\vec{E}$ are computed as follows:

$$\vec{B} = 2 \cdot \Re \vec{d_1} \tag{5.7}$$

$$\vec{E} = 2 \vec{h} \cdot \Re \vec{d_2} - \vec{h} \tag{5.8}$$

$$\vec{h} = 5 - (\text{Iteration} * (5 / \text{Iter}_{\max})) \tag{5.9}$$

where Iter_{max} = maximum iterations assigned, $\vec{h}$ = a linearly decreasing value from 5 to 0 in steps of 0.1, $\Re\vec{d_1}$ and $\Re\vec{d_2}$ = random $[0, 1]$ number.

In this figure, (A, B) are the hyena which will adjust its location toward the prey (A^*, B^*) based on the values of Eqs. (5.7)−(5.9). More information on encircling can be found in Reference [25].

5.3.2.2.2 Hunting

During this phase, the hyena pack will move close to the prey and proceed to the attack. This process is diagrammatically shown in Fig. 5.4.

This phase is represented with Eqs. (5.10)−(5.12);

$$\vec{D}_h = \left| \vec{B} \cdot \vec{P}_h - \vec{P}_k \right| \tag{5.10}$$

$$\vec{P}_k = \vec{P}_h - \vec{E} \cdot \vec{D}_h \tag{5.11}$$

$$\vec{C}_h = \vec{P_k} + \vec{P_{k+1}} + \cdots + \vec{P_{k+N}} \tag{5.12}$$

where $\vec{P}_h$ = leader which moves closer to prey and $\vec{P_k}$ = positions of other hyenas in the pack, and N = total hyenas in the pack. More information can be found in Reference [25].

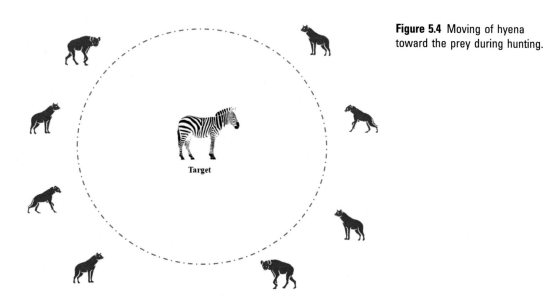

Figure 5.4 Moving of hyena toward the prey during hunting.

5.3.2.2.3 Attacking

When the hyena moves to attack the prey, other hyenas in the group follow the same technique and the group attack will result in the prey being killed. When the prey is dead, every hyena in the pack is on or closer to the prey. This process is the convergence of the chosen agents toward the optimal location. Fig. 5.5 depicts the process graphically and the mathematical expression of this process is shown in Eq. (5.13).

$$\vec{P}(x+1) = \frac{\overrightarrow{C_h}}{N} \tag{5.13}$$

where $\vec{P}(x+1)$ is the best position, in which every hyena in the pack converges. In this work, the SHA is initiated with the following parameters: number of hyenas (agents) $= N = 30$, search dimension $(D) = 3$, Iter$_{max} = 5000$, and stopping criteria $=$ maximization of Kapur/Otsu function or Iter$_{max}$. The proposed SMA is used in finding the best threshold for the HMRI under assessment.

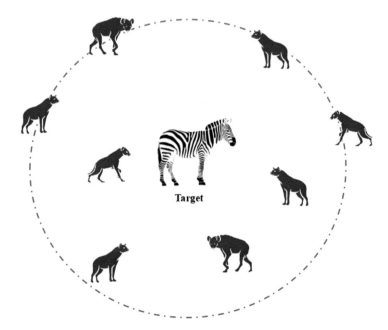

Figure 5.5 Attacking and killing of the prey by the hyena pack.

Target

5.3.3 Postprocessing

This practice is employed to segment the infected fragments in HMRI from the preprocessed picture.

The mathematical expression of LS is depicted in Eqs. (5.14)–(5.16).

$$\text{OP} = \frac{\partial\phi(F, V)}{\partial V} \tag{5.14}$$

where, ϕ = curve vector with spatial constraint (F) and sequential variable (V), O = speed utility, and P = innermost curve normal vector ϕ. The curve growth can be realized as LS segmentation by inserting the active contour $\phi(F, V)$ as the beginning bounding-box (BB).

The innermost usual vector P is signified as:

$$P = \frac{-\nabla\varphi}{\nabla\varphi} \tag{5.15}$$

where ∇ = gradient function.

The final LS growth is articulated as:

$$\frac{\partial\varphi}{\partial V} = O|\nabla\varphi| \tag{5.16}$$

More details on the LS can be found in References [26–28].

5.3.4 Performance evaluation

The proposed HMRI examination system consists of a preprocessing phase and a postprocessing phase. In order to confirm the quality of the processed image, the performances of these two phases are evaluated separately.

The performance of the preprocessing phase computes the image quality values by comparing the threshold image with the original image and the expressions considered in this task are discussed in Chapter 3 (subsection 3.2.12). These values are separately computed for SHA + KE, SHA + FE and SHA + BCV and, based on these values, the performance of the preprocessing task is confirmed.

The abnormal section in HMRI is then extracted using the LS and, to confirm the segmentation performance, a pixel-level comparison of the GT and extracted section is performed as discussed in Chapter 4 (subsection 4.3.4). The computed measures are then evaluated separately to confirm the performance of SHA + KE + LS, SHA + FE + LS and SHA + BCV + LS.

5.4 Results and discussion

This section demonstrates the experimental results achieved with the HMRI of HVSMR2016, With all the results obtained using MATLAB.

Initially, the axial-view HMRI was considered for the appraisal and methods, such as SHA + KE, SHA + FE and SHA + BCV-supported pre-processing, were initially applied to improve the visibility of the abnormal HMRI section.

When the SHA + KE was initiated with the assigned SHA parameters, it helped to provide the preprocessed image with optimally selected threshold, as shown in Fig. 5.6. This figure confirms that the threshold is separated into three sections, the background, normal pixels and the pixels belongs to the abnormal section. Fig. 5.7 shows the convergence of the SHA + KE at the maximized value of KE, and this convergence confirms that the proposed technique converged for iteration <1500 for a chosen $\text{Iter}_{\text{max}} = 5000$. This also confirms that the SHA + KE helped to achieve a faster convergence compared to the other threshold selection procedures employed herein. This process was repeated for other images used in this work and similar results were obtained.

Fig. 5.8 presents the threshold outcome of the SHA + KE for the chosen test images and in this image the abnormal area is

Figure 5.6 Optimally assigned threshold using SHA + KE.

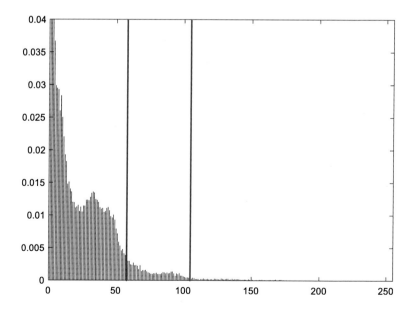

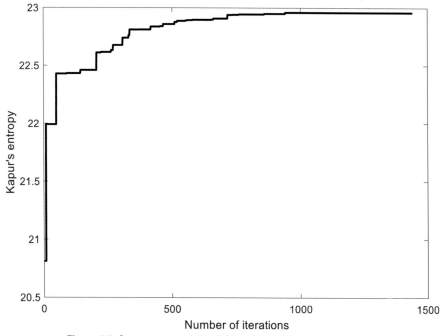

Figure 5.7 Convergence of the SHA + KE search at maximized KE.

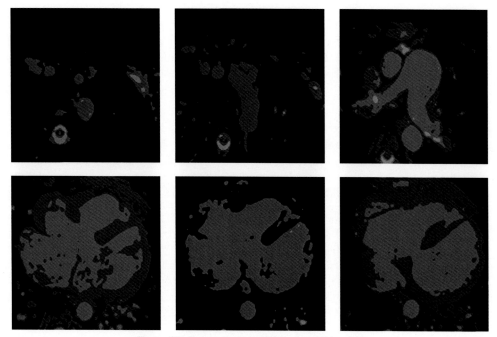

Figure 5.8 Preprocessed HMRI using SHA + KE.

clearly visible compared to the normal area and background. The enhanced area can be extracted using the LS. The thresholded HMRI is compared with the original image and the necessary image quality values (IQV) are computed. These values are presented in Table 5.1. The average IQV values shown in the table confirm that the proposed scheme provides better values of NAE, PSNR, and SSIM. A pictorial representation of the SSIM is presented in Fig. 5.9 which also confirms the superiority of the proposed scheme.

The thresholding result obtained with SHA + FE is shown in Fig. 5.10 for the chosen axial-view HMRI and the computed IQN are presented in Table 5.2. From the average values computed in the table, it can be noted that the IQV of SHA + FE is poor compared to the SHA + KE and this result confirms that the FE is not suitable for preprocessing of the HMRI.

A similar process is then repeated with SHA + BCV and the obtained results are presented in Fig. 5.11. The IQV achieved with the pixel-level comparison of the preprocessed and original image is depicted in Table 5.3. This table verifies that the IQV obtained is better than that for the SHA + FE and similar to that for the SHA + KE. The overall comparison of the chosen IQV is depicted in Fig. 5.12, which confirms that the thresholding with Kapur and Otsu is better than with the FE.

Finally, the LS segmentation is implemented on the preprocessed image and the extracted image (binary form) is presented in Fig. 5.13. When compared to the GT, the extracted images provide a better pixel match. Later, a comparison of extracted HMRI with the GT was implemented and the necessary quality measures (QM) computed.

Table 5.1 Image measures obtained for SHA + KE thresholded HMRI.

MRI slice	NAE	RMSE	PSNR	FSIM	NCC	AD	SC	SSIM
Im1	0.8063	24.4579	20.3624	0.7069	0.3569	18.2463	3.5588	0.2436
Im2	0.7516	23.8527	20.5801	0.7517	0.4059	18.4319	3.5780	0.2435
Im3	0.4364	19.8142	22.1913	0.7623	0.7126	16.1651	1.7415	0.2435
Im4	0.3141	20.0664	22.0814	0.7269	0.7509	16.7572	1.6546	0.5320
Im5	0.5010	33.3319	17.6736	0.6975	0.6084	27.2132	2.0296	0.3404
Im6	0.3070	18.6798	22.7034	0.7510	0.7552	15.8032	1.6575	0.5655
Average	0.5194	23.3671	20.9320	0.7327	0.5983	18.7695	2.37	0.3614

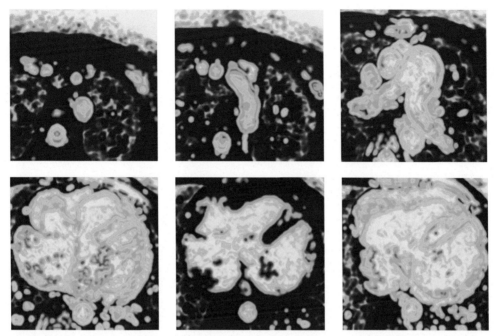

Figure 5.9 Graphical depiction of the structural similarity between the original and threshold images.

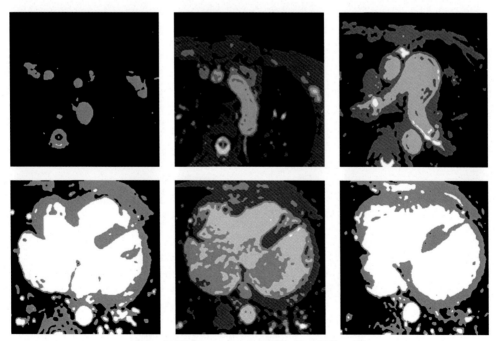

Figure 5.10 Preprocessed HMRI using SHA + FE.

Table 5.2 Image measures obtained for SHA + FE thresholded HMRI.

MRI slice	NAE	RMSE	PSNR	FSIM	NCC	AD	SC
Im1	0.8947	27.0797	19.4779	0.6894	0.5709	15.7950	1.1947
Im2	0.5778	19.7769	22.2076	0.7587	1.3390	−2.0161	0.5001
Im3	0.7103	35.6695	17.0849	0.6735	1.5170	−12.1436	0.3973
Im4	1.7389	113.1812	7.0553	0.6428	2.7030	−83.2358	0.1304
Im5	0.7501	52.0398	13.8041	0.6946	1.7412	−33.5662	0.3170
Im6	1.8126	114.6396	6.9441	0.6548	2.7801	−84.8908	0.1240
Average	1.0807	60.3978	14.4289	0.6856	1.7752	−33.3429	0.4439

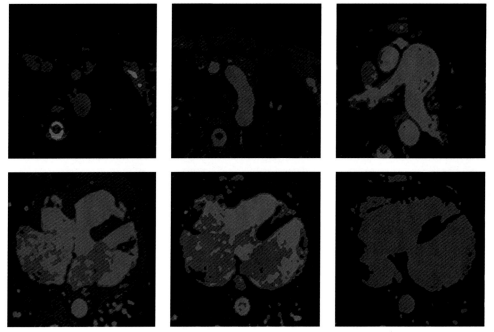

Figure 5.11 Preprocessed HMRI using SHA + BCV.

Fig. 5.14 presents the confusion matrix constructed for a sample HMRI, and this image confirms that the SHA + KE + LS helps to obtain a segmentation accuracy of 99.87% on the chosen image. Similar results were achieved for other images using the KE, FE, and BCV, and the obtained results are presented in Tables 5.4 and 5.5. These tables confirm that the developed HMRI

Table 5.3 Image measures obtained for Otsu thresholded HMRI.

MRI slice	NAE	RMSE	PSNR	FSIM	NCC	AD	SC
Im1	0.3593	11.7665	26.7179	0.8284	0.6779	8.1317	2.0535
Im2	0.3659	12.0413	26.5173	0.8219	0.7137	8.9735	1.8297
Im3	0.3177	14.7297	24.7670	0.8039	0.7695	11.7655	1.6091
Im4	0.2567	16.3379	23.8669	0.7836	0.8086	13.6941	1.4594
Im5	0.2615	16.8140	23.6174	0.7732	0.7979	14.2017	1.5015
Im6	0.3502	21.2656	21.5773	0.7633	0.6851	18.0236	2.0378
Average	0.3185	15.4925	24.5106	0.7957	0.7421	12.4650	1.7485

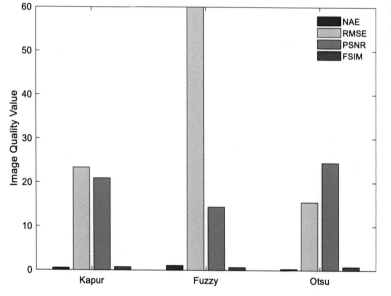

Figure 5.12 Performance evaluation of preprocessed HMRI.

examination system works well on the chosen images. The overall result achieved for the considered test images of the HVSMR2016 benchmark data set is presented in Fig. 5.15, and this comparison confirms that the SHA + KE + LS helped to achieve a better QM compared to the other procedures. These results confirm its clinical significance and the proposed method can be used to examine real HMRI images collected from hospitals.

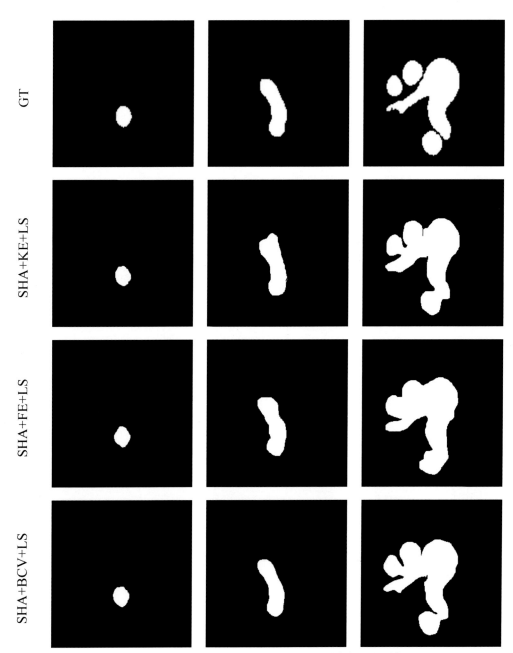

Figure 5.13 Segmented abnormal section from HMRI using LS.

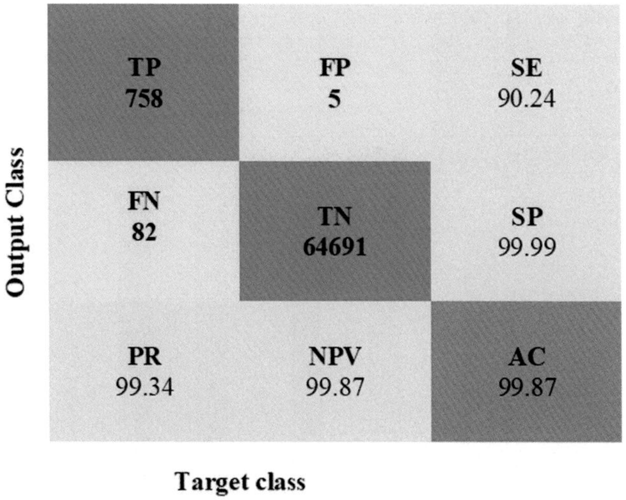

Figure 5.14 Confusion matrix obtained when comparing segmented HMRI with GT.

Table 5.4 Initial quality measures achieved when comparing segmented HMRI with GT.

Approach	Image	TP	FN	TN	FP	Jaccard	Dice
Kapur + LS	Im1	758	82	64,691	5	89.7041	94.5727
	Im2	2716	64	62,279	476	83.4152	90.9578
	Im3	774	30	53,672	1197	38.6807	55.7838
Fuzzy + LS	Im1	758	82	64,681	15	88.6550	93.9864
	Im2	2702	78	62,340	416	84.5432	91.6243
	Im3	767	37	53,357	1512	33.1174	49.7567
Otsu + LS	Im1	766	74	64,661	35	87.5429	93.3577
	Im2	2640	140	62,704	52	93.2203	96.4912
	Im3	789	15	53,857	1012	43.4471	60.5758

Table 5.5 Quality measures computed for various segmented images.

Approach	Image	AC	PR	SE	SP	NPV
Kapur + LS	Im1	99.8672	99.3447	90.2381	99.9923	99.8734
	Im2	99.1760	85.0877	97.6978	99.2415	99.8973
	Im3	97.7961	39.2694	96.2687	97.8184	99.9441
Fuzzy + LS	Im1	99.8520	98.0595	90.2381	99.9768	99.8734
	Im2	99.2462	86.6581	97.1942	99.3371	99.8750
	Im3	97.2177	33.6551	95.3980	97.2443	99.9307
Otsu + LS	Im1	99.8337	95.6305	91.1905	99.9459	99.8857
	Im2	99.7070	98.0684	94.9640	99.9171	99.7772
	Im3	98.1553	43.8090	98.1343	98.1556	99.9722

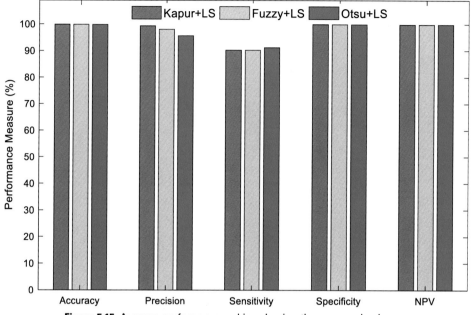

Figure 5.15 Average performance achieved using the proposed scheme.

5.5 Conclusion

Biosignal- and bioimage-supported cardiac abnormality detection are widely employed in hospitals, with the image-based method giving more complete information about the heart than the bio-signals. Heart MRI (HMRI) is a commonly adopted imaging scheme to detect and diagnose various heart diseases. In this work, a joint thresholding and segmentation process was employed to detect abnormalities in HMRI collected from the HVSMR2016 benchmark data set. The thresholding was implemented using the SHA and KE/FE/BCV a trilevel thresholding was employed to enhance the images. After the enhancement, the affected area was mined using the LS. The performances of thresholding and segmentation procedures are separately confirmed by computing the necessary quality measures and, based on these values, the performance of the proposed system was verified. The experimental results of this study confirm that the thresholding outcome of SHA + KE is better than for SHA + FE and SHA + BCV. The segmentation result by all considered approaches was similar, and this method extracts the abnormal area in the HMRI with better accuracy. In future, the proposed scheme could be implemented to examine real patient HMRIs collected from hospitals.

References

[1] Dey N, Ashour AS, Shi F, Fong SJ, Sherratt RS. Developing residential wireless sensor networks for ECG healthcare monitoring. IEEE Transactions on Consumer Electronics 2017;63(4):442–9.

[2] Mukhopadhyay S, Biswas S, Roy AB, Dey N. Wavelet based QRS complex detection of ECG signal. arXiv; 2012. preprint arXiv:1209.1563.

[3] Nandi S, Roy S, Dansana J, Ben W, Karaa A, Ray R, et al. Cellular automata based encrypted ECG-hash code generation: an application in inter-human biometric authentication system. International Journal of Computer Network and Information Security 2014;6(11).

[4] Lin H, Rajinikanth V. Development of softcomputing tool to evaluate heart MRI slices. International Journal of Computer Theory and Engineering 2019;11(5):80–3.

[5] Arunmozhi S, Lin H, Rajinikanth V. Examination of 2D cardiac MRI using softcomputing assisted scheme. In: 2019 IEEE international conference on system, computation, automation and networking (ICSCAN). IEEE; March 2019. p. 1–4.

[6] Fernandes SL, Rajinikanth V, Kadry S. A hybrid framework to evaluate breast abnormality using infrared thermal images. IEEE Consumer Electronics Magazine 2019;8(5):31–6.

[7] Zuluaga MA, Bhatia K, Kainz B, Moghari MH, Pace DF, editors. Reconstruction, segmentation, and analysis of medical images: first international workshops, RAMBO 2016 and HVSMR 2016, held in conjunction with MICCAI 2016, Athens, Greece, October 17, 2016, revised selected papers, vol. 10129. Springer; 2017.

[8] Yu L, Cheng JZ, Dou Q, Yang X, Chen H, Qin J, Heng PA. Automatic 3D cardiovascular MR segmentation with densely-connected volumetric convnets. In: International conference on medical image computing and computer-assisted intervention. Cham: Springer; September 2017. p. 287–95.

[9] Yu L, Yang X, Qin J, Heng PA. 3D FractalNet: dense volumetric segmentation for cardiovascular MRI volumes. In: Reconstruction, segmentation, and analysis of medical images. Cham: Springer; 2016. p. 103–10.

[10] Guttman MA, Zerhouni EA, McVeigh ER. Analysis of cardiac function from MR images. IEEE Computer Graphics and Applications 1997;17(1):30–8.

[11] Backhaus SJ, Metschies G, Billing M, Kowallick JT, Gertz RJ, Lapinskas T, Schuster A. Cardiovascular magnetic resonance imaging feature tracking: impact of training on observer performance and reproducibility. PLoS One 2019;14(1):e0210127.

[12] Khan AH, Verma R, Bajpai A, Mackey-Bojack S. Unusual case of congestive heart failure: cardiac magnetic resonance imaging and histopathologic findings in cobalt cardiomyopathy. Circulation: Cardiovascular Imaging 2015;8(6):e003352.

[13] Friedrich MG. Steps and leaps on the path toward simpler and faster cardiac MRI scanning. 2021.

[14] Attili AK, Schuster A, Nagel E, Reiber JH, van der Geest RJ. Quantification in cardiac MRI: advances in image acquisition and processing. The International Journal of Cardiovascular Imaging 2010;26(1):27–40.

[15] Gupta A, Gulati GS, Seth S, Sharma S. Cardiac MRI in restrictive cardiomyopathy. Clinical Radiology 2012;67(2):95–105.

[16] Cheng HD, Chen YH, Sun Y. A novel fuzzy entropy approach to image enhancement and thresholding. Signal Processing 1999;75(3):277—301.

[17] Sarkar S, Paul S, Burman R, Das S, Chaudhuri SS. A fuzzy entropy based multi-level image thresholding using differential evolution. In: International conference on swarm, evolutionary, and memetic computing. Cham: Springer; December 2014. p. 386—95.

[18] Roopini IT, Vasanthi M, Rajinikanth V, Rekha M, Sangeetha M. Segmentation of tumor from brain MRI using fuzzy entropy and distance regularised level set. In: Computational signal processing and analysis. Singapore: Springer; 2018. p. 297—304.

[19] Rajinikanth V, Satapathy SC. Segmentation of ischemic stroke lesion in brain MRI based on social group optimization and Fuzzy-Tsallis entropy. Arabian Journal for Science and Engineering 2018;43(8):4365—78.

[20] Otsu N. A threshold selection method from gray-level histograms. IEEE Transactions on Systems, Man, and Cybernetics 1979;9(1):62—6.

[21] Dhiman G, Kumar V. Spotted hyena optimizer: a novel bio-inspired based metaheuristic technique for engineering applications. Advances in Engineering Software 2017;114:48—70.

[22] Dhiman G, Kaur A. Spotted hyena optimizer for solving engineering design problems. In: 2017 international conference on machine learning and data science (MLDS). IEEE; December 2017. p. 114—9.

[23] Dhiman G, Kumar V. Multi-objective spotted hyena optimizer: a multi-objective optimization algorithm for engineering problems. Knowledge-Based Systems 2018;150:175—97.

[24] Dhiman G, Kumar V. Spotted hyena optimizer for solving complex and non-linear constrained engineering problems. In: Harmony search and nature inspired optimization algorithms. Singapore: Springer; 2019. p. 857—67.

[25] Dhiman G, Kaur A. Optimizing the design of airfoil and optical buffer problems using spotted hyena optimizer. Designs 2018;2(3):28.

[26] Li C, Xu C, Gui C, Fox MD. Distance regularized level set evolution and its application to image segmentation. IEEE Transactions on Image Processing 2010;19(12):3243—54.

[27] Rajinikanth V, Fernandes SL, Bhushan B, Sunder NR. Segmentation and analysis of brain tumor using Tsallis entropy and regularised level set. In: Proceedings of 2nd international conference on micro-electronics, electromagnetics and telecommunications. Singapore: Springer; 2018. p. 313—21.

[28] Rajinikanth V, Satapathy SC, Dey N, Fernandes SL, Manic KS. Skin melanoma assessment using Kapur's entropy and level set—a study with bat algorithm. In: Smart intelligent computing and applications. Singapore: Springer; 2019. p. 193—202.

6

CNN-based segmentation of brain tumor from T2-weighted MRI slices

6.1 Introduction

In the current era, the illness rate in humans is steadily increasing due to multiple causes. Infections in the inner body organs are believed to be more sensitive than infections in the exterior organs. Although there are a number of vital internal organs in human physiology, the brain is considered to be the primary organ, which controls all other parts of the body. The main roles of the brain are information processing, decision-making, and storing the vital information. Abnormalities in the brain will affect its basic operations and untreated irregularities will lead to various problems, including disability and/or death. Brain tumor is a severe abnormality and the leading cause of problems in the central nervous system (CNS); therefore timely recognition and treatment can prevent disability and/or death.

Based on its orientation and severity, tumors are classified into different categories, with each category needing its own treatment procedures including medication, radiotherapy, and/or surgery, as discussed in Refs. [1−3]. The most commonly encountered brain tumors include low-grade glioma (LGG), high-grade glioma (HGG), and Glioblastoma multiforme (GBM), and specific treatment procedures are necessary to control and remove these tumors. LGG starts in glial cells and can severely impact the performance of the CNS. GBM is also a severe condition that arises in brain/spinal cord and can cause various related problems [4−6].

Detection of tumors is generally achieved by radiological procedures, with MRI being one of the most commonly adopted imaging schemes before computed tomography (CT). After recording the MRI, a personal check or computerized diagnosis is performed to detect the location and severity of the cancer. After the class of tumor has been identified by MRI, the doctor

Magnetic Resonance Imaging: Recording, Reconstruction and Assessment. https://doi.org/10.1016/B978-0-12-823401-3.00005-5

will plan and implement the necessary treatment procedure. In the literature, a number of image examination schemes have been developed and employed to appraise the tumor in MRI slices using a chosen modality. The MRI axial view is extensively adopted by doctors and researchers due to its simplicity.

The hospital-level detection of tumors using the selected MRI slice is a difficult task when a group screening is scheduled and the classification of tumors into various classes is also essential during the detection, assessment, and treatment. The development of an accurate computerized detection scheme (CDS) is necessary to extract and evaluate the tumor and, in this work, a convolutional neural network (CNN)-based procedure is implemented.

The experimental study is performed with the clinical-grade brain MRIs of the BRATS2015, TCIA-LGG, and TCIA-GNM image data sets. These data sets contains 3D brain MRIs of varied modalities and, in this research work, the T2-weighted modality alone is considered for the evaluation. Initially, a 3D to 2D exchange is performed with ITK-Snap [7,8]. After the 2D slice extraction, each image is resized to $224 \times 224 \times 3$ pixels and the converted image is then used for the experimental evaluation. The BRATS2015 data set provides brain MRIs without the skull section and also provides the ground truth (GT) image for examination. The TCIA data set has the brain MRI with the skull and no GT is provided. In this work, 500 MRI slices were used from each image case ($500 \times 4 = 2000$ images) in which 400 images were used to train the CNN scheme and 100 images were used to validate the segmentation performance of the CNN.

In this research, well-known CNN segmentation schemes, such as UNet, SegNet, VGG-UNet, and VGG-SegNet were employed to mine the cancerous region in the MRI slice with improved accuracy. After extracting the tumor, it is compared with the GT and the required quality measures (QM) are calculated and, based on the accomplished QM, the merit of the implemented CNN segmentation procedure is validated. Further, the merit of CNN is also authenticated using the traditional segmentation routines, in which thresholding is employed using the spotted hyena algorithm and Shannon's entropy (SHA + SE) and a semi/automated segmentation process. In this work, relative appraisals of procedures, like watershed (WS), level set (LS), active contour (AC) and Chan-Vese (CV) are also presented.

The proposed experimental investigation is realized with MATLAB and the average result of this work proves that the general performance of VGG-SegNet is better than the other methods considered in this chapter.

Of the remainder of this chapter is presented as follows: Section 6.2 presents the previous research works, Section 6.3 shows the methodology, and Sections 6.4 and 6.5 give the results and conclusion, respectively.

6.2 Related research

Because of their importance, a number of tumor segmentation and examination methods have been developed by the researchers with the help of traditional and modern methods. The modern methods, such as soft-computing-supported and CNN-based approaches, are widely employed in the literature due to the improved performance.

Soft-computing (SC)-assisted tumor recognition proposes a method using the following stages: image resizing, tumor enhancement using trilevel thresholding, and segmentation. Although the steps involved in this process are more, the computation effort needed is much less and in most cases, it helps to obtain better segmentation for the MRI with varied modalities.

Along with the SC schemes, CNN-supported tumor extraction is also widely implemented in research to support better detection. The computation effort needed in a CNN scheme is comparatively higher than for SC schemes and each approach needs to be trained for the image data set in order to make the algorithm learn and distinguish the tumor pixels from the other image pixels. Usually, CNN segmentation employs an encoder–decoder section to extract the tumor from the chosen bioimage. Table 6.1 summarizes some chosen image segmentation procedures employed to extract the abnormal section from the brain MRI slices.

A detailed review of brain tumor detection can be found in Ref. [26] and this work confirms the need for a suitable tumor extraction procedure to mine the abnormal areas with better accuracy. In this research, a detailed assessment of the chosen CNN schemes and SC-based tumor segmentation is presented and discussed.

6.3 Methodology

This section demonstrates the methodology employed to extract and evaluate tumor segments from MRI slices. The block diagram of this work is presented in Fig. 6.1. This work implements the CNN-based and soft-computing-supported segmentation procedure to obtain the growth with enhanced accuracy.

Table 6.1 Summary of CNN-supported brain tumor extraction methods.

Reference	Methodology executed
Iqbal et al. [9]	Mining and assessment of brain tumors in 2D MRI slices of BRATS2015 are presented, and this work employed deep CNN to segment the cancer section with improved accuracy
Pereira et al. [10]	A CNN-supported segmentation scheme is implemented to mine gliomas in brain MRI slices. This work used the images from the BRATS2013 and BRATS2015 databases
Thaha et al. [11]	This work employed the bat algorithm with optimized and enhanced CNN to mine tumors from brain MRI slices
Havaei et al. [12]	This work implemented deep neural networks-assisted automated segmentation and categorization of brain cancer
Kermi et al. [13]	This research employed UNet-supported segmentation of brain lumps from multimodal MRI slices
Xu et al. [14]	Long-/short-term-memory (LSTM)-based UNet is executed to accurately obtain the lump parts from MRI slices
Lachinov et al. [15]	This work proposed a cascaded UNet scheme to extract the glioma region from MRI slices
Daimary et al. [16]	Segmentation of the tumor region from the MRI slice is executed using various CNN schemes
Alqazzaz et al. [17]	SegNet-supported mining of the growth part from multimodal MRI is discussed with experimental results
Nema et al. [18]	This work developed generative adversarial network (GAN)-based RescueNet to extract the tumors in BR images
Kadry et al. [19]	This work implemented a UNet-based scheme to extract the stroke lesions in MRI slices of varied modalities.
Ghassemi et al. [20]	Implementation of pretrained deep GAN is used to discuss the classification of MRI slices with tumor
Pravitasari et al. [21]	Segmentation of an abnormal section in brain MRI is implemented using pretrained VGG16-UNet
Akkus et al. [22]	A thorough evaluation of deep-learning-assisted brain tumor assessment using MRI is discussed
Gordillo et al. [23]	This work presents a comprehensive survey of methods employed to mine tumors in brain MRIs
Pereira et al. [10]	CNN-supported automatic segmentation of tumors in brain MRIs is presented
Wadhwa et al. [24]	A comprehensive review of automated and semiautomated brain tumor mining from MRI slices is presented and discussed
Pereira et al. [25]	Implementation of a semantic segmentation system to mine the abnormal fragment in brain MRIs is discussed using an adaptive feature recombination technique

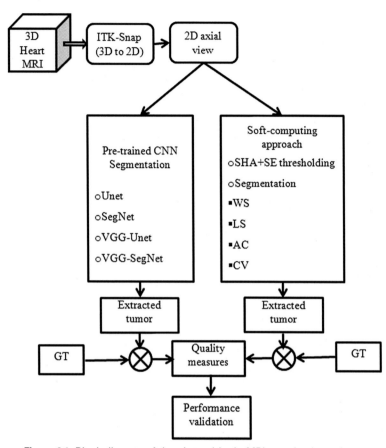

Figure 6.1 Block diagram of the planned brain MRI examination scheme.

Initially, 2D slices are extracted from the chosen MRI using ITK-Snap and then resized to $224 \times 224 \times 3$ pixels. This image is then separately segmented using the CNN scheme and using the proposed soft-computing scheme. The attained tumor part is then compared with the GT and, based on the obtained QM, identification of the best segmentation process is confirmed.

6.3.1 Brain MRI database

Usually, the proposed image examination scheme must be tested and validated using medically ranked MRI slices. In this work, the MRIs were collected from benchmark data sets, including BRATS2015 (LGG and HGG) [27], TCIA-LGG [28], and TCIA-HGG [29,30], and from each database 500 images (RGB

Table 6.2 Images considered in this research work.

Database	Class	Dimension	Training images	Validation images
BRATS	LGG	$224 \times 224 \times 3$	500	100
	HGG	$224 \times 224 \times 3$	500	100
TCIA	LGG	$224 \times 224 \times 3$	500	100
	GBM	$224 \times 224 \times 3$	500	100

scale) were extracted and then resized. Table 6.2 presents the information regarding the images considered in this research work.

Figs. 6.2 and 6.3 present the sample trial imagery used in this work. The BRATS images are found without the artifact (skull)

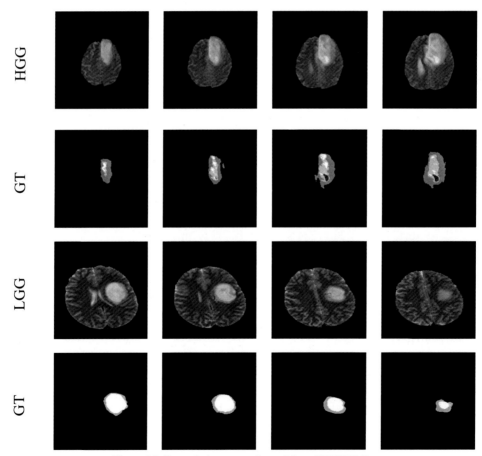

Figure 6.2 Example images collected from BRATS2015.

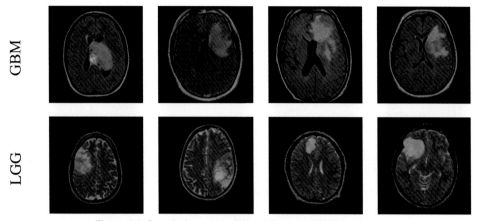

Figure 6.3 Sample images collected from the TCIA database.

and the TCIA imagery are obtainable with the skull fragment. All the images were processed based on the scheme depicted in Fig. 6.1.

6.3.2 CNN segmentation

Due to their advantages and enhanced accuracy, CNN segmentation schemes are widely employed to extract the chosen area from general and medical images. The main limitation of CNN segmentation is that it has to learn for the chosen test images. When this scheme is completely trained to obtain the necessary division from the 2D image (gray/RGB scale) then it will offer a more accuracy than the other existing methods described in the literature. The CNN-based segmentation procedures, like UNet [31] and SegNet [32], are very popular among researchers, and a number of earlier works have confirmed the performance of these schemes on a chosen image database [19−21]. Normally, CNN segmentation consists of the encoder−decoder section, which helps to extract the relevant area using the SoftMax layer after the final layer of the decoder section. The performance of these schemes is also enhanced by replacing the encoder section with well-known pretrained schemes, such as the VGG and ResNet, and in this work, UNet and SegNet enhanced using the VGG to obtain the cancer areas from MRI slices.

6.3.2.1 VGG-UNet

The traditional UNet was invented in 2015 [31] and was originally developed to mine for the essential division of biomedical images. Due to its advantageous properties, a number of segmentation problems are solved using the traditional UNet scheme

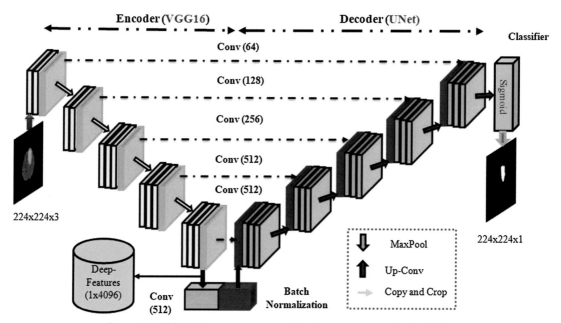

Figure 6.4 VGG-UNet scheme employed to mine tumors from brain MRIs.

and, to enhance its performance, VGG-UNet was developed in 2018 using the VGG11 scheme [33,34]. Recently, a number of modifications have been employed in this scheme to improve the encoder and decoder sections and, in the proposed work, VGG11 is replaced with VGG16 and the design of the structure is depicted in Fig. 6.4. The encoder section consists of the traditional VGG16 and the decoder segment is associated with the UNet, and this combination assists in extracting the cancerous area with better accuracy. Initially, this system was trained using GT, BRATS images, and TCIA images and finally, 100 images from each case were used to validate the performance of this scheme. Other related information on VGG-UNet is found in Ref. [35].

6.3.2.2 VGG-SegNet

The structure of VGG-SegNet depicted in Fig. 6.5 is the enhanced version of the SegNet proposed in 2015. This structure also consists of an encoder–decoder section in which the downsampling and up-sampling are continuously performed until the SoftMax layer extracts the tumor. The implementation and training process for this scheme is similar to that for the VGG-UNet and essential pretraining is necessary to use this procedure for brain MRI data. Recent works implemented with this scheme

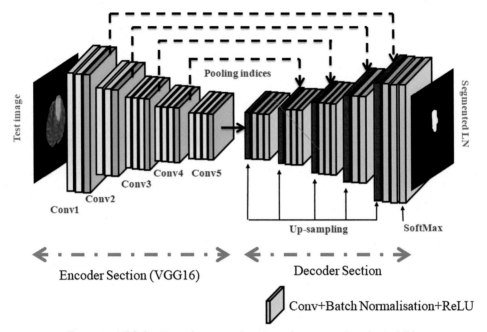

Figure 6.5 VGG-SegNet scheme employed to mine tumors from brain MRIs.

can be found in Ref. [36]. In this work, the pretrained architecture is considered to extract the tumor section from the T2-weighted MRI slices of VRATS and TCIA.

6.3.2.3 Joint thresholding and segmentation

To validate the performance of the proposed CNN scheme, this work also employed the soft-computing-based tumor segmentation process. This work implements a joint thresholding and segmentation procedure to extract the tumor from the MRI.

The necessary thresholding is implemented with a trilevel scheme using the spotted hyena algorithm and Shannon's entropy (SHA + SE). This practice is related to the techniques discussed in Chapter 5. The thresholding procedure is employed to separate the MRI slice into tumor, normal, and background regions. After this separation, the growth part is extracted by the semi/automated segmentation procedures discussed in Chapters 4 and 5. The segmentation procedures, such as WS, LS, AC, and CV, are employed to extract the tumor.

After mining the tumor region, a comparison is performed with the GT and the necessary QM are computed. The obtained QM of this scheme are then compared with the QM achieved with the CNN and, based on these values, its significance in brain MRI assessment is confirmed. The soft-computing-based procedures are less complexes than the CNN schemes and can be easily implemented to obtain the necessary outcome.

6.4 Results and discussion

This part of the chapter presents the experimental results attained using the workstation, Intel i5 2.9 GHz processor with 12 GB RAM and 4 GB VRAM equipped with MATLAB.

Initially, pretuning of the CNN scheme is performed using the training images of BRATS and TCIA ($400 \times 4 = 1600$ images) until the CNN scheme extracts the tumor section. After training the CNN system to perfection, the BRATS images are initially used to test the performance of the system and the results attained from various layers of the VGG-SegNet are presented for demonstration, as in Fig. 6.6.

The outcome attained while transferring the information from the encoder to the decoder section is presented in Fig. 6.6. This figure depicts the test image used, the intermediate layer results (MaxPool), and the extracted tumor with the SoftMax layer. After extracting the tumor, it is then converted into a binary image (background = 0 and tumor = 1 pixels). From this figure, it can be seen that the CNN helps to get a better result. A similar

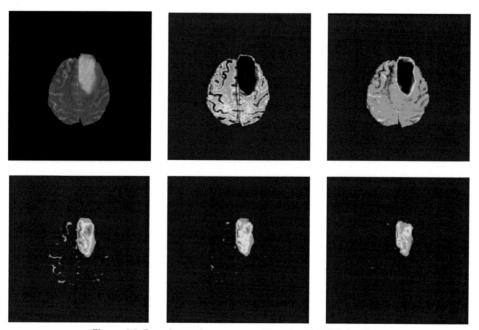

Figure 6.6 Experimental outcome achieved with VGG-SegNet.

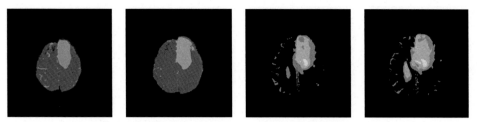

Figure 6.7 Thresholding outcome obtained with SHA + SE.

procedure is repeated with UNet, SegNet, and VGG-UNet, and the extracted tumors are separately stored for further assessment.

The soft-computing-supported tumor extraction is then implemented on the chosen brain MRI and the results are compared with the CNN result. In this work, a trilevel thresholding is implemented using SHA + SE to enhance the tumor section and, after the enhancement, it is extracted using the chosen segmentation procedure. Fig. 6.7 presents the enhanced MRI slices with SHA + SE and this confirms that the proposed scheme groups the MRI pixels into tumor, normal tissue, and background efficiently for the chosen T2-weighted image.

After enhancement, the tumor region is then extracted using the WS scheme; various phase outcomes are presented in Fig. 6.8. WS is an automated method, which helps to get the

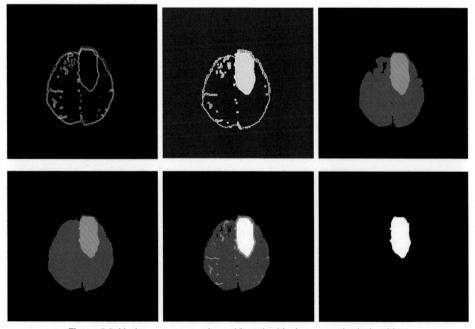

Figure 6.8 Various stage results achieved with the watershed algorithm.

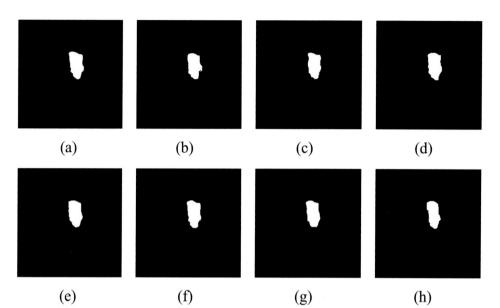

Figure 6.9 Extracted tumor section using CNN and soft-computing techniques: (A) UNet; (B) SegNet; (C) VGG-UNet; (D) VGG-SegNet; (E) WS; (F) LS; (G) AC; (H) CV.

required area using procedures, such as edge detection, watershed fill, morphological operation, and mining. Segmentation of the tumor in the thresholded image is also performed using methods, such as LS, AC, and CV, and the extracted area using each technique is presented in Fig. 6.9. In this image, Fig. 6.9A–H present the binary versions of the tumor extracted from the CNN and soft-computing schemes.

A relative assessment among the binary forms of tumor image and GT is then performed to compute the necessary QM to confirm its practicability. During this process, a pixel-level assessment between the GT and tumor segment is executed and the achieved results are depicted in Tables 6.3 and 6.4. Table 6.3 presents the pixel-related information, like TP, FN, TN, and FP and computed values of the Jaccard and Dice. This table confirms that the QM attained with VGG-SegNet is better than that for the other methods and the results by SHA + SE + WS achieve second position and help to get the best results compared with the other methods in this study.

Table 6.4 presents other QM, such as accuracy (AC), precision (PR), sensitivity (SE), specificity (SP), and negative predictive value (NPV) achieved during this work and this also confirms that VGG-SegNet provides the best result. an individual comparison of the computed QM is then performed and the pictorial representation

Table 6.3 Initial image quality measures attained for a sample image.

Approach	TP	FN	TN	FP	Jaccard	Dice
UNet	1657	101	63,621	157	86.5274	92.7772
SegNet	1568	48	63,674	246	84.2105	91.4286
VGG-UNet	1532	47	63,675	282	82.3213	90.3036
VGG-SegNet	1702	46	63,676	112	91.5054	95.5643
SHA + SE + WS	1673	39	63,683	141	90.2860	94.8951
SHA + SE + LS	1637	90	63,632	177	85.9769	92.4598
SHA + SE + AC	1621	30	63,692	193	87.9067	93.5642
SHA + SE + CV	1553	9	63,713	261	85.1892	92.0024

Table 6.4 Quality measures obtained with CNN and soft-computing methods.

Approach	AC	PR	SE	SP	NPV
UNet	99.6063	91.3451	94.2548	99.7538	99.8415
SegNet	99.5514	86.4388	97.0297	99.6151	99.9247
VGG-UNet	99.4980	84.4542	97.0234	99.5591	99.9262
VGG-SegNet	99.7589	93.8258	97.3684	99.8244	99.9278
SHA + SE + WS	99.7253	92.2271	97.7220	99.7791	99.9388
SHA + SE + LS	99.5926	90.2426	94.7887	99.7226	99.8588
SHA + SE + AC	99.6597	89.3605	98.1829	99.6979	99.9529
SHA + SE + CV	99.5880	85.6119	99.4238	99.5920	99.9859

of these comparisons can be found in Fig. 6.10. Fig. 6.10A and B present the Jaccard and Dice and Fig. 6.10C—G present the AC, PR, SE, S,P and NPV values, respectively, for a chosen MRI slice. This confirms that VGG-SegNet helps to obtain a better outcome (Fig. 6.11).

Figs. 6.12—6.14 present the results achieved for the additional test imagery used in the assessment. These figures also confirm that the QM achieved in VGG-SegNet for HGG, LGG, and GBM images are superior to those of other CNN and soft-computing methods, and this confirms that the VGG-SegNet provided better segmentation results for the considered data set and so, in the future, this approach could be used to examine clinical-grade brain MRI images of varied modalities, such as flair, T1, T1C, and DW.

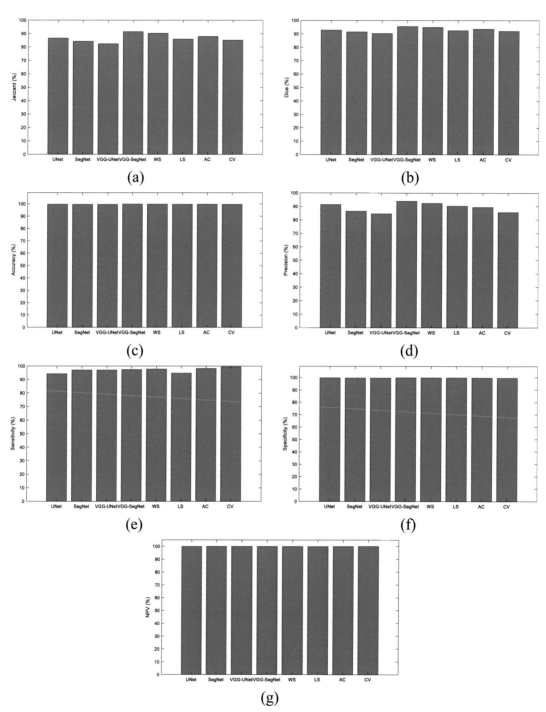

Figure 6.10 Comparing the attained quality measures graphically.

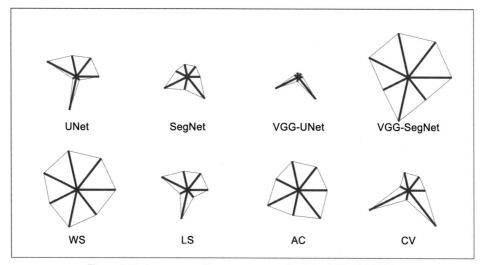

Figure 6.11 Average quality measures attained for LGG of BRATS.

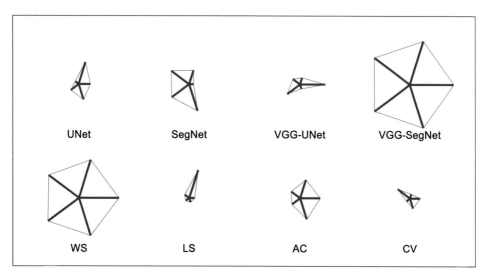

Figure 6.12 Average quality measures attained for HGG of BRATS.

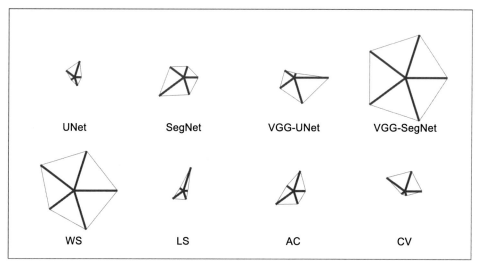

Figure 6.13 Average quality measures attained for TCIA-LGG.

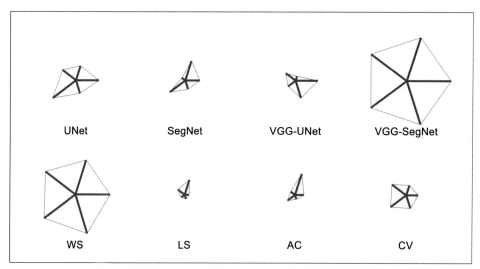

Figure 6.14 Average quality measures attained for TCIA-GBM.

6.5 Conclusion

Bioimage-supported assessment of brain abnormalities is an approved procedure in hospitals to accurately detect the location and severity of the disease. Brain tumor detection using MRI is a common procedure and a number of assessment procedures have been already proposed and discussed by researchers. This work proposed the CNN-based segmentation of tumor sections in brain MRI slices and the performance of this method is confirmed with a comparative estimation linking the existing CNN schemes and the soft-computing methods. Initially, the segmentation task was implemented using UNet, SegNet, VGG-UNet, and VGG-SegNet. Later, this task was implemented using SHA + SE-based thresholding and WS/LS/AC/CV segmentation. After obtaining the growth, a relative appraisal was executed between the tumor and GT and essential QM such as Jaccard, Dice, AC, PR, SE, SP, and BPV were calculated. The experimental outcome of this work confirms that the overall performance of VGG-SegNet is better for the test images collected from BRATS and TCIA data sets. In the future, this method could be employed to examine clinical-grade MRIs collected from hospitals.

References

[1] Fernandes SL, Tanik UJ, Rajinikanth V, Karthik KA. A reliable framework for accurate brain image examination and treatment planning based on early diagnosis support for clinicians. Neural Computing and Applications 2020; 32(20):15897—908.

[2] Pugalenthi R, Rajakumar MP, Ramya J, Rajinikanth V. Evaluation and classification of the brain tumor MRI using machine learning technique. Journal of Control Engineering and Applied Informatics 2019;21(4):12—21.

[3] Dey N, Rajinikanth V, Shi F, Tavares JMR, Moraru L, Karthik KA. Social-Group-Optimization based tumor evaluation tool for clinical brain MRI of Flair/diffusion-weighted modality. Biocybernetics and Biomedical Engineering 2019;39(3):843—56.

[4] Zacharaki EI, Wang S, Chawla S, Soo Yoo D, Wolf R, Melhem ER, Davatzikos C. Classification of brain tumor type and grade using MRI texture and shape in a machine learning scheme. Magnetic Resonance in Medicine: An Official Journal of the International Society for Magnetic Resonance in Medicine 2009;62(6):1609—18.

[5] Sultan HH, Salem NM, Al-Atabany W. Multi-classification of brain tumor images using deep neural network. IEEE Access 2019;7:69215—25.

[6] Smoll NR, Schaller K, Gautschi OP. Long-term survival of patients with glioblastoma multiforme (GBM). Journal of Clinical Neuroscience 2013; 20(5):670—5.

[7] http://www.itksnap.org/pmwiki/pmwiki.php.

[8] Yushkevich PA, Piven J, Hazlett HC, Smith RG, Ho S, Gee JC, Gerig G. User-guided 3D active contour segmentation of anatomical structures:

significantly improved efficiency and reliability. Neuroimage 2006;31(3): 1116−28.

[9] Iqbal S, Ghani MU, Saba T, Rehman A. Brain tumor segmentation in multispectral MRI using convolutional neural networks (CNN). Microscopy Research and Technique 2018;81(4):419−27.

[10] Pereira S, Pinto A, Alves V, Silva CA. Brain tumor segmentation using convolutional neural networks in MRI images. IEEE Transactions on Medical Imaging 2016;35(5):1240−51.

[11] Thaha MM, Kumar KPM, Murugan BS, Dhanasekeran S, Vijayakarthick P, Selvi AS. Brain tumor segmentation using convolutional neural networks in MRI images. Journal of Medical Systems 2019;43(9):1−10.

[12] Havaei M, Davy A, Warde-Farley D, Biard A, Courville A, Bengio Y, et al. Brain tumor segmentation with deep neural networks. Medical Image Analysis 2017;35:18−31.

[13] Kermi A, Mahmoudi I, Khadir MT. Deep convolutional neural networks using U-Net for automatic brain tumor segmentation in multimodal MRI volumes. In: International MICCAI brainlesion workshop. Cham: Springer; September 2018. p. 37−48.

[14] Xu F, Ma H, Sun J, Wu R, Liu X, Kong Y. July). LSTM multi-modal UNet for brain tumor segmentation. In: 2019 IEEE 4th international conference on image, vision and computing (ICIVC). IEEE; 2019. p. 236−40.

[15] Lachinov D, Vasiliev E, Turlapov V. Glioma segmentation with cascaded UNet. In: International MICCAI brainlesion workshop. Cham: Springer; September 2018. p. 189−98.

[16] Daimary D, Bora MB, Amitab K, Kandar D. Brain tumor segmentation from MRI images using hybrid convolutional neural networks. Procedia Computer Science 2020;167:2419−28.

[17] Alqazzaz S, Sun X, Yang X, Nokes L. Automated brain tumor segmentation on multi-modal MR image using SegNet. Computational Visual Media 2019; 5(2):209−19.

[18] Nema S, Dudhane A, Murala S, Naidu S. RescueNet: an unpaired GAN for brain tumor segmentation. Biomedical Signal Processing and Control 2020; 55:101641.

[19] Kadry S, Damaševičius R, Taniar D, Rajinikanth V, Lawal IA. U-net supported segmentation of ischemic-stroke-lesion from brain MRI slices. In: 2021 seventh international conference on bio signals, images, and instrumentation (ICBSII). IEEE; March 2021. p. 1−5.

[20] Ghassemi N, Shoeibi A, Rouhani M. Deep neural network with generative adversarial networks pre-training for brain tumor classification based on MR images. Biomedical Signal Processing and Control 2020;57:101678.

[21] Pravitasari AA, Iriawan N, Almuhayar M, Azmi T, Fithriasari K, Purnami SW, Ferriastuti W. UNet-VGG16 with transfer learning for MRI-based brain tumor segmentation. Telkomnika 2020;18(3):1310−8.

[22] Akkus Z, Galimzianova A, Hoogi A, Rubin DL, Erickson BJ. Deep learning for brain MRI segmentation: state of the art and future directions. Journal of digital imaging 2017;30(4):449−59.

[23] Gordillo N, Montseny E, Sobrevilla P. State of the art survey on MRI brain tumor segmentation. Magnetic Resonance Imaging 2013;31(8):1426−38.

[24] Wadhwa A, Bhardwaj A, Verma VS. A review on brain tumor segmentation of MRI images. Magnetic Resonance Imaging 2019;61:247−59.

[25] Pereira S, Alves V, Silva CA. Adaptive feature recombination and recalibration for semantic segmentation: application to brain tumor segmentation in MRI. In: International conference on medical image

computing and computer-assisted intervention. Cham: Springer; September 2018. p. 706—14.

[26] Nazir M, Shakil S, Khurshid K. Role of deep learning in brain tumor detection and classification (2015—2020): a review. Computerized Medical Imaging and Graphics; 2021. p. 101940.

[27] Menze BH, Jakab A, Bauer S, Kalpathy-Cramer J, Farahani K, Kirby J. The multimodal brain tumor image segmentation benchmark (BRATS). IEEE Transactions on Medical Imaging 2014;34(10):1993—2024.

[28] Pedano N, Flanders AE, Scarpace L, Mikkelsen T, Eschbacher JM, Hermes B. Radiology data from the cancer genome atlas low grade glioma [TCGA-LGG] collection. In: The cancer imaging archive; 2016. https://doi.org/10.7937/K9/TCIA.2016.L4LTD3TK.

[29] Scarpace L, Mikkelsen T, Cha S, Rao S, Tekchandani S, Gutman D, Saltz JH, Erickson BJ, Pedano N, Flanders AE, Barnholtz-Sloan J, Ostrom Q, Barboriak D, Pierce LJ. Radiology data from the cancer genome atlas glioblastoma multiforme [TCGA-GBM] collection [data set]. In: The cancer imaging archive; 2016. https://doi.org/10.7937/K9/TCIA.2016.RNYFUYE9.

[30] Clark K, Vendt B, Smith K, Freymann J, Kirby J, Koppel P, Moore S, Phillips S, Maffitt D, Pringle M, Tarbox L, Prior F. The cancer imaging archive (TCIA): maintaining and operating a public information repository. Journal of Digital Imaging 2013;26(6):1045—57. https://doi.org/10.1007/s10278-013-9622-7.

[31] Ronneberger O, Fischer P, Brox T. U-net: convolutional networks for biomedical image segmentation. In: International conference on medical image computing and computer-assisted intervention. Cham: Springer; October 2015. p. 234—41.

[32] Badrinarayanan V, Kendall A, Cipolla R. Segnet: a deep convolutional encoder-decoder architecture for image segmentation. IEEE Transactions on Pattern Analysis and Machine Intelligence 2017;39(12):2481—95.

[33] Iglovikov V, Shvets A. Ternausnet: U-net with vgg11 encoder pre-trained on imagenet for image segmentation. arXiv 2018. preprint arXiv:1801.05746.

[34] Shi J, Dang J, Cui M, Zuo R, Shimizu K, Tsunoda A, Suzuki Y. Improvement of damage segmentation based on pixel-level data balance using VGG-unet. Applied Sciences 2021;11(2):518.

[35] Rajinikanth V, Kadry S. Development of a framework for preserving the disease-evidence-information to support efficient disease diagnosis. International Journal of Data Warehousing and Mining (IJDWM) 2021;17(2):63—84.

[36] Kadry S, Taniar D, Damaševičius R, Rajinikanth V, Lawal IA. Extraction of abnormal skin lesion from dermoscopy image using VGG-SegNet. In: 2021 seventh international conference on bio signals, images, and instrumentation (ICBSII). IEEE; March 2021. p. 1—5.

Automated detection of ischemic stroke with brain MRI using machine learning and deep learning features

7.1 Introduction

The current advances in the healthcare area facilitate enhancement of the hospital environment and, due to this, a number of modern disease diagnosis facilities are employed.

Although hospitals are capable of having the modern amenities, the procedures, such as disease diagnosis, treatment, and patient recovery remain challenging tasks. Recently, disease occurrence in humans has been gradually rising, causing a large diagnostic burden to healthcare workers, who executes a number of prescribed actions to identify the illness and its severity in the patient. To decrease this analytical burden, an initial level of screening is performed using a suitable computer algorithm and its result/report is then evaluated by a doctor.

The usual practice in the current scenario is as follows: patients with sensitive illness wish to visit specialty hospitals to obtain suitable analysis and treatment. Modern hospitals are capable of diagnosing disease with computerized exploratory schemes to ensure appropriate and precise disease identification [1,2]. These computerized procedures also have the capacity to classify the disease condition into phases, such as mild, modest, and severe; enabling patients who need instant treatment to be identified.

The existing modern facilities and computerized healthcare information gathering and maintenance facilities support the execution of automatic illness analytical facilities. The current enhancement in artificial intelligence (AI) procedures, such as machine learning (ML) and deep learning (DL) have helped to improve the illness detection accuracy for sensitive diseases arising in different organs including the brain [3−5].

In the brain, stroke is one of the frequent illnesses, and the chief cause of stroke is a reduced/irregular blood supply. Stroke is classified as either ischemic stroke (IS) or hemorrhagic stroke (HS) [6,7]. When the brain segment does not obtain sufficient blood/oxygen, it leads to stroke, which may create a temporary/permanent disability, with untreated stroke leading to death. Compared to HS, the incidence rate of IS is higher; therefore a considerable number of IS detection methods have been suggested by researchers [8–10].

Biosignal (electroencephalogram)-based stroke recognition requires complex preprocessing and computation practice because of its nonlinear nature. The previous works on stroke analysis verified that the bioimage (MRI)-based method offers essential information in contrast to the biosignal-based methods. The MRI-based diagnosis can be executed using 3D or 2D images of modalities, such as T1, flair, and DW. Moreover, MRI-based detection provides the necessary information, such as severity, location, and volume of the affected brain segment, which plays a major part in the treatment planning and execution.

This chapter aims to suggest a DL structure to support segmentation and classification of the IS section from 2D MRI slices. It consists of the following phases to accomplish improved illness recognition; (1) segmentation of IS using VGG-UNet, (2) mining of features, like Hu, gray level co-occurrence matrix (GLCM), and local binary pattern (LBP), (3) deep-features extraction, (4) feature ranking and serial concatenation, and (5) classification and validation.

In this work, the essential test pictures of normal/stroke class were collected from the ISLES2015 challenge data set. The ISLES2015 data set contain the 3D brain MRIs of T1, flair, and DW modalities without the skull region and each volunteer's picture is connected with two ground-truth (GT) images. In the proposed research, the diagnosis performance of VGG-UNet is initially verified on the considered image data set and the segmented image is considered to extract the features, such as Hu and GLCM.

The test images are then treated using LBP and the extracted features are combined with Hu and GLCM to obtain the ML feature vector. The disease detection performance of DL and DL + ML is individually scrutinized using the binary classifiers. The results of this work are then examined and based on the computed values of the quality measures (QM) the performance of the implemented IS detection scheme is confirmed. Herein, the proposed work is executed using the axial-view MRI slices and, during this examination, the flair MRI modality is considered. Every modality image is individually examined with the proposed method and the results presented.

The other part of this research is arranged as follows: Section 7.2 presents the earlier related works, Section 7.3 presents the methodology, and Sections 7.4–7.6 present the experimental results, discussion, and conclusion, respectively.

7.2 Earlier research

Recently, a number of MRI-supported IS detection procedures were proposed and discussed because of their clinical significance. Research by Johnson et al. [11] verified that cerebrovascular accidents (stroke) are the second main cause of death and third major cause of disability. This work also showed that the incidence of IS is steadily increasing in low- and middle-income countries. In addition, the article substantiated that >85% of stroke-caused deaths and disabilities occur in low- and middle-income countries.

The previous investigation also proved that MRI-based IS recognition is widely suggested for precisely identifying the position and severity of stroke. When a stroke is identified, the doctor will arrange and handle the recovery of the patient. Earlier literature projected a substantial amount of AI-based IS lesion segmentation and classification procedures.

Table 7.1 fives an outline of modern stroke segmentation and classification methods described in the literature for the ISLES2015 database.

The work of Zhang et al. [20] presented a detailed review of the identification of various IS assessment events planned in the literature. This work also recommended the usage of DL-supported segmentation and classification to reduce the diagnostic burden of stroke detection. Based on the recommendations of earlier research, a detailed examination of IS in the brain MRI is examined using DL and DL + ML schemes.

7.3 Methodology

The performance of an automated medical image examination scheme depends on the methodology employed. In this work, the DL and ML schemes are implemented to improve the IS detection in a flair modality MRI slice. The proposed scheme is depicted in Fig. 7.1. Initially, the 2D slices are collected from the ISLES2015 database using ITK-Snap and the extracted images are resized to $224 \times 224 \times 3$ pixels. The image features (texture

Table 7.1 Summary of the preferred IS detection methods.

References	Stroke detection methodology
Maier et al. [12]	This work presents a comprehensive evaluation regarding the segmentation of IS from multispectral MRI using a support vector machine
Maier et al. [13]	A complete assessment of semiautomated/automated segmentation events for IS detection is discussed using the flair, T1, and DW modalities of ISLES2015
Maier et al. [14]	Extraction of IS from the selected MRI modality is implemented by means of an extra tree forest algorithm
Maier et al. [15]	This research offers a comprehensive discussion of the consideration of diverse MRI modalities to record IS and its inspection techniques
Subbanna et al. [16]	This work presents the segmentation of IS in flair MRI by means of a customized Markov random field technique
Zhang et al. [17]	Multiplane information blending-based mining of IS is presented with a variety of MRI modalities
Kumar et al. [18]	Execution of a modified deep learning system for IS extraction is demonstrated
Rajinikanth and Satapathy [19]	A comprehensive investigation into thresholding and segmentation-based IS mining is presented with different MRI modalities
Lin et al. [10]	The requirement for the joint thresholding/segmentation system to extract IS in an MRI slice is discussed and the outcomes of flair and DW modalities are presented
Hemanth et al. [9]	The role of multimodality image fusion in enhancing the detection accuracy of IS is demonstrated using a soft computing system

and shape) are then separately extracted using the chosen ML and DL schemes. In this work, the VGG-UNet is employed to obtain the IS lesion from the chosen test image. The extracted IS lesion is in a binary form and from this image, the ML features, like Hu and GLCM, are obtained. Later, the LBP features are also separately extracted with various weight values (W = 1 to 4), and then combined with the Hu and GLCM to get the 1D ML feature vector. The VGG-UNet section consists of the encoder and decoder unit, and the output of the encoder section will offer the deep features of dimensions $1 \times 1 \times 4096$ and, after the possible dropout, the feature size is reduced to $1 \times 1 \times 1024$ and this feature vector is then considered to train, test, and validate the binary classifier. The outcome of this framework is separately recorded and assessed. Further, this DL feature vector is then combined with the ML feature vector using a sorting and serial concatenation operation (DL + ML). The combined feature vector is then considered to classify the brain MRI slices using the binary classifier. In this work, a fivefold cross-validation is used and the outcomes are recorded.

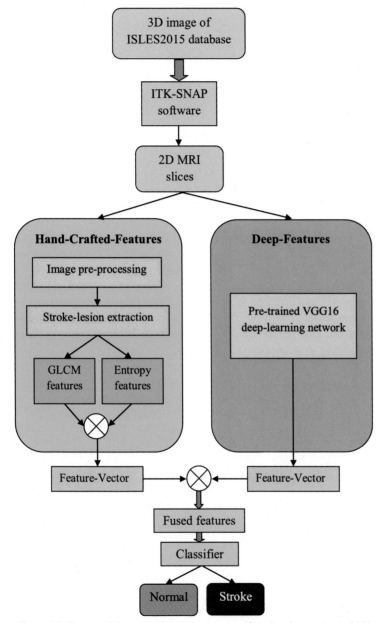

Figure 7.1 Proposed framework to examine the IS lesion from a brain MRI.

7.3.1 Image database

The essential test image (flair modality MRI) considered in this work is obtained from the ISLES2015 database [12]. This data set is one of the most widely analyzed MRI databases in the literature,

Table 7.2 Number of test images in this work.

Image type	Dimension	Images		
		Total	**Training**	**Testing**
Normal	224 × 224 × 3	1000	700	300
Stroke	224 × 224 × 3	1000	700	300

which consists of the brain MRI of flair, T1, and DW modalities along with two ground-truth images. This data set is available without the skull section and, to reduce the computation complexity, 2D MRI slices are extracted and considered. The ITK-Snap is used to extract the 2D slices from the 3D MRI and the extracted images are available with dimensions of $77 \times 77 \times 3$ pixels and, to implement the VGG-UNet, all these images are modified to $224 \times 224 \times 3$ pixels and the pictures are then examined. The total numbers of images used in this research for normal and stroke classes are shown in Table 7.2 and the sample imagery is presented in Fig. 7.2.

7.3.2 VGG16 architecture

In the literature, a number of DL schemes are executed to scrutinize a class of signals and images. Compared to other

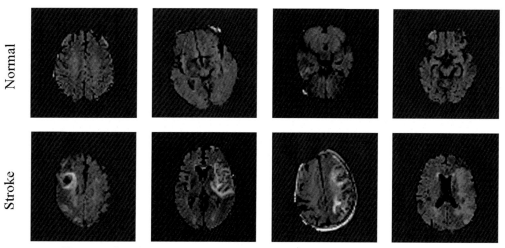

Figure 7.2 Sample test images of the flair modality considered in this research.

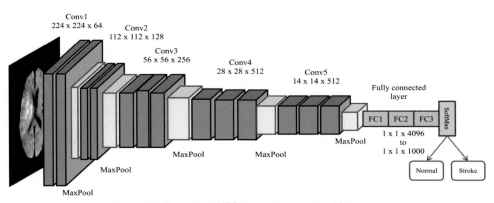

Figure 7.3 Pretrained VGG16 used to examine IS lesions.

methods, the VGG16 architecture seems to be a simple and effective method to inspect a class of images (gray/RGB scale). Fig. 7.3 presents the structure of the VGG16 with various sections. This structure has 16 working layers (convolutional + fully connected) excluding the maximum pooling layer. This using has five convolutional layer sections and three fully connected (FC) sections, with a 50% dropout rate. The final section of the fully connected section (FC3) is linked with the classifier unit, which helps to segregate the image into normal/stroke sections.

The necessary information about the VGG16 can be found in Refs. [21,22]. The maximum pooling section transfers the essential information from one convolution section to the other by considering only the maximized value of features, and this task has been already discussed in earlier works. The final convolution layer presents a 1D feature vector of dimensions $1 \times 1 \times 4096$ and after the possible dropout features are decreased to $1 \times 1 \times 1024$, which is then used for testing the classifier.

The VGG16 combined with UNet (VGG-UNet) is also used to obtain the IS lesion in the MRI slice and the related information on this operation is presented in Chapter 6 (subsection 6.3.2.1).

7.3.3 Feature extraction

The performance of the automated disease classification system depends mainly on the features extracted and used to drive the classifier unit. In this work, two different features are used to examine the IS lesion detection performance of the proposed system.

Initially, the detection of the IS from flair modality MRI is achieved using the pretrained VGG16 with the DL features, and the performance of this system is confirmed using various binary classifiers, such as SoftMax, naïve Bayes (NB), random forest (RF), K nearest-neighbor (KNN), and SVM with linear kernel [23,24]. After recording the performance using a fivefold cross-validation process, this work is repeated with the DL + ML-based features.

In this work, the ML features are extracted from the binary image of the IS lesion image and the LBP-treated image and this information can be found in Refs. [25−27].

The features used in this work are depicted in Eqs. (7.1)−(7.9);

$$DL_{(1 \times 1024)} = VGG16_{(1,1)}, VGG16_{(1,2)}, \ldots, VGGgg16_{(1,1024)} \tag{7.1}$$

$$GLCM_{(1 \times 25)} = GLCM_{(1,1)}, GLCM_{(1,2)}, \ldots, GLCM_{(1,25)} \tag{7.2}$$

$$Hu_{(1 \times 3)} = Hu_{(1,1)}, Hu_{(1,2)}, Hu_{(1,3)} \tag{7.3}$$

$$LBP_{w1(1 \times 59)} = LBP_{(1,1)}, LBP_{(1,2)}, \ldots, LBP_{(1,59)} \tag{7.4}$$

$$LBP_{w2(1 \times 59)} = LBP_{(1,1)}, LBP_{(1,2)}, \ldots, LBP_{(1,59)} \tag{7.5}$$

$$LBP_{w3(1 \times 59)} = LBP_{(1,1)}, LBP_{(1,2)}, \ldots, LBP_{(1,59)} \tag{7.6}$$

$$LBP_{w4(1 \times 59)} = LBP_{(1,1)}, LBP_{(1,2)}, \ldots, LBP_{(1,59)} \tag{7.7}$$

$$ML_{(1 \times 264)} = GLCM_{(1 \times 25)} + Hu_{(1 \times 3)} + LBP_{w1(1 \times 59)} + \cdots + LBP_{w4(1 \times 59)} \tag{7.8}$$

$$DL + ML_{(1 \times 776)} = DF_{50\%(1 \times 512)} + ML_{(1 \times 264)} \tag{7.9}$$

Eq. (7.1) presents the DL features obtained with the VGG16. Eqs. (7.2) and (7.3) present the GLCM and Hu moments obtained from the binary image of the IS lesion extracted using the VGG-UNet. Eqs. (7.4)−(7.7) present the LBP features obtained from the LBP-enhanced images for a chosen weight of W = 1−4 and the essential details on the LBP image can be found in Ref. [28]. Eq. (7.8) depicts the ML features considered in this work and Eq. (7.9) presents the DL + ML features. The DL features of Eq. (7.9) are obtained using 50% dropout in Eq. (7.1) and hence the DL + ML (concatenated feature) helped to get a 1D feature subset with a value of $1 \times 1 \times 776$. The classification task is once again repeated using the DL + ML and the attained results are recorded and compared.

7.3.4 Performance evaluation

The performance of the proposed IS examination system is verified using the test images (300 numbers) of the normal/stroke classes. During the classification task, a fivefold cross-validation is implemented and the attained values in the confusion matrix (CM) are then recorded to confirm the performance of the proposed scheme. The CM helps to obtain the necessary information, such as true-positive (TP), false-negative (FN), true-negative (TN) and false-positive (FP) values initially. From these values, other measures, such as accuracy (AC), precision (PR), sensitivity (SE), specificity (SP), and negative predictive value (NPV) are derived. These values are then considered to confirm the classifier performance with DL and DL + ML features. The necessary information regarding the performance measures considered in this research can be found in Refs. [29,30].

7.4 Experimental results

This division of the chapter reveals the investigational result achieved via a workstation with an Intel i5 2.9 GHz processor with 12 GB RAM and 4 GB VRAM set with MATLAB.

Initially, the pretuning of the VGG-UNet is executed with the training images of ISLES2015 until this scheme segments the IS lesion accurately. The segmentation superiority of VGG-UNet is verified using the testing images. When the proposed task is implemented, the SoftMax section existing in the decoder section helps to extract the IS lesion and provides a binary picture in which the IS lesion is allocated with a pixel value of one (1) and the surroundings are allocated with zero (0). The output layer of the encoder section is linked with fully connected layers to extract the DL features with dimensions of $1 \times 1 \times 1024$ (see Fig. 7.3).

The experimental outcome achieved with VGG-UNet for a sample test image is depicted in Fig. 7.4. Fig. 7.4A presents the test image, Fig. 7.4B−E depict the intermediate layer outcomes, and Fig. 7.4F shows the binary image extracted with the SoftMax layer. This image is then used to extort the Hu and GLCM features.

The sample LBP images for a chosen weight W = 1−4 are presented in Fig. 7.5 for both the MRI normal and stroke classes and each image helps to get a feature value of $1 \times 1 \times 59$.

VGG16-based image classification with DL features was initially performed on the considered IS lesion MRI images with the flair modality. During the pretraining task, 700 images (normal/stroke classes) were used to train the network using the actual and augmented images. The in-built augmentation process rotates the images in clockwise and counterclockwise directions with an angle from 0 to 80 degrees (in steps of 20 degrees) and the augmented image is then considered to train the VGG16 for the chosen data set. After a twofold training

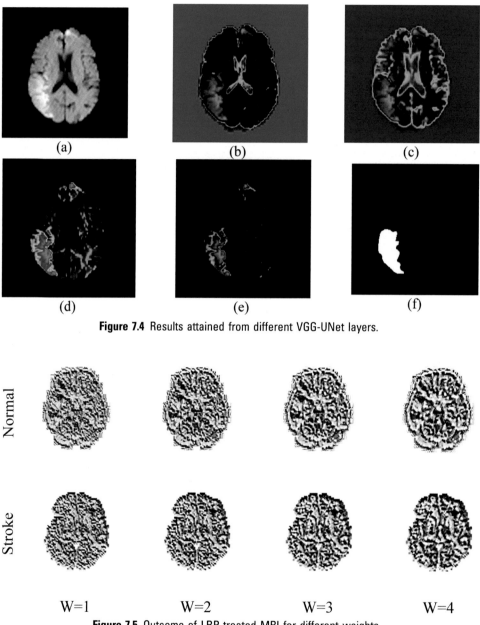

Figure 7.4 Results attained from different VGG-UNet layers.

Figure 7.5 Outcome of LBP-treated MRI for different weights.

process, the 300 test images (normal/stroke classes) are considered to test the performance of the VGG16.

The intermediate images extracted during the training process are depicted in Fig. 7.6. Fig. 7.6A presents the result of the

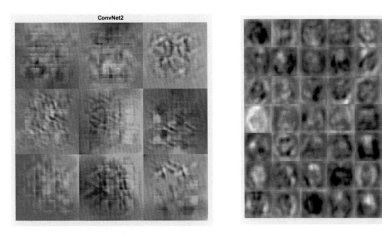

(a)   (b)

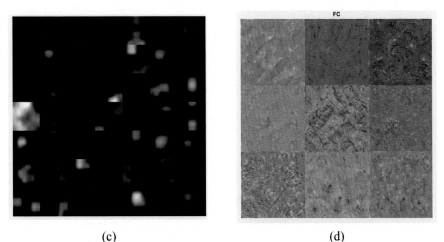

(c)   (d)

Figure 7.6 Intermediate-layer results of VGG16 during the training process: (A) convolution net representation; (B) ConvNet image output; (C) MaxPooling; (D) fully connected representation.

convolutional layer 2 (ConvNet2), Figs. 7.6B and C present the ConvNet3 outcome and MaxPool images, respectively. Fig. 7.6D presents the image features available at the fully connected layer. After completely training the VGG16, its performance was then tested using the test images of normal/stroke classes. Fig. 7.7 presents the outputs of the various layers of VGG16 during the testing operation and for better visibility a hot-color map is used to better demonstrate the images. Figs. 7.7A–F present the various layer outcomes, such as ConvNet and MaxPool. Similar results can be achieved for other images used in this research work.

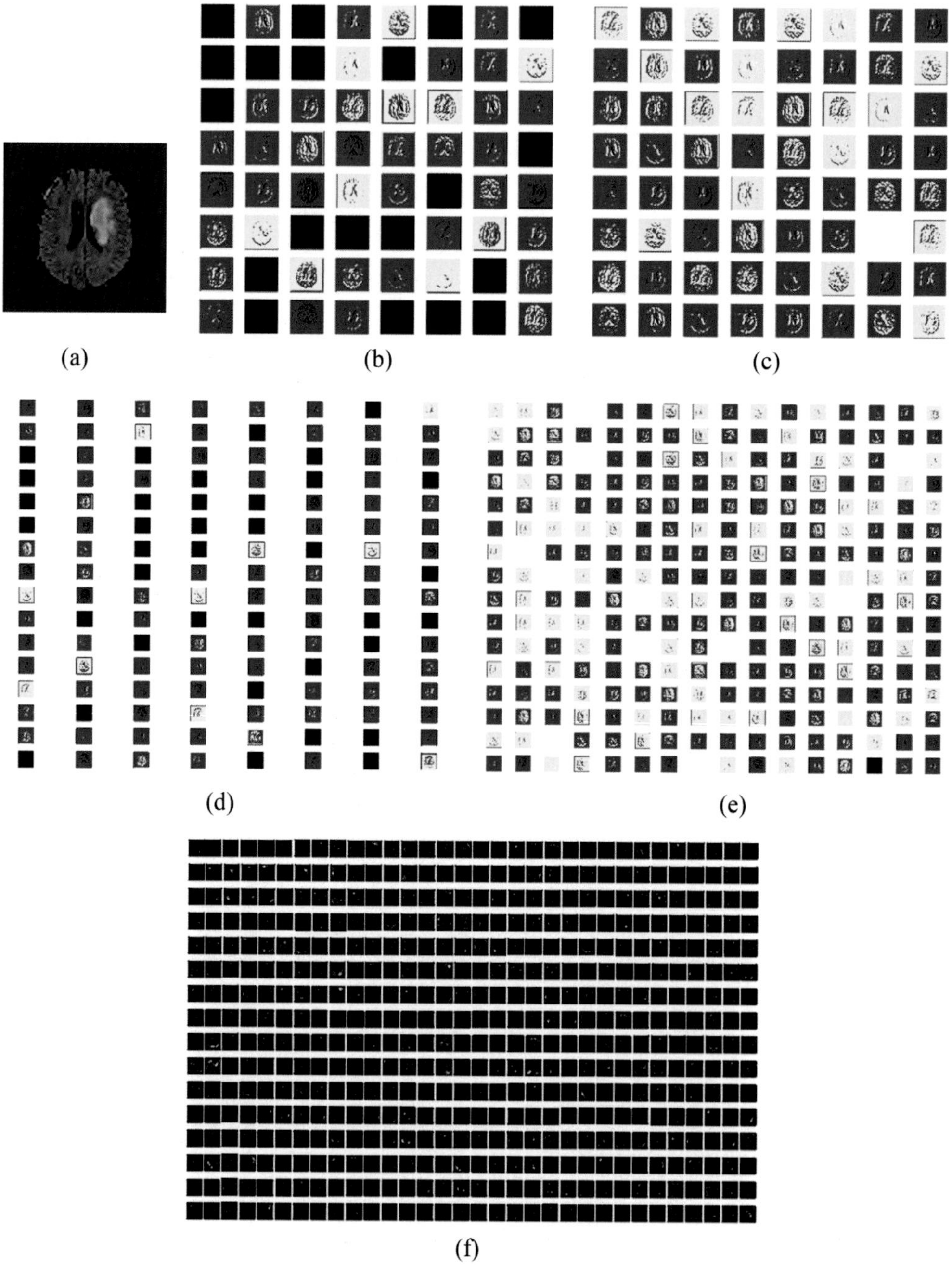

Figure 7.7 Sample results obtained during the testing process: (A) test image; (B) Conv output ($8 \times 8 = 64$); (C) MaxPool output ($8 \times 8 = 64$); (D) Conv output ($16 \times 8 = 128$); (E) MaxPool output ($16 \times 16 = 256$); (F) MaxPool output ($16 \times 32 = 512$).

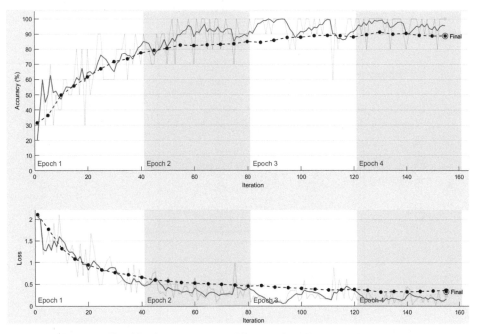

Figure 7.8 Classification accuracy and loss attained for the testing operation.

During this operation, the number of epoch was assigned as 50, the iteration size was fixed as 2000, and the error value was allocated as $1e^{-5}$. The training and testing value is continued until the error value of the loss function is $\leq 1e^{-5}$.

The classification task is initiated with VGG16 using a fivefold cross-validation and the best result achieved among the five trials was chosen as the final result. The convergences attained during the classification task with SoftMax classifier and the corresponding CM are presented in Figs. 7.8 and 7.9, respectively. From this CM, it can be seen that the classification accuracy achieved for SoftMax with DL features is 91.7%, and other QMs are also obtainable from the CM.

The VGG16-supported disease detection results achieved for the sample test images are presented in Fig. 7.10 and these results substantiate that this system performs well for flair modality MRI slices.

The classifier outcomes for all five trials are presented in Table 7.3 and these results confirm that trial 4 provides better accuracy compared to the other trials. The Glyph-Plot presented in Fig. 7.11 also confirms that the overall result of trial 4 is better than the other trials and hence, the trial 4 value is selected as the

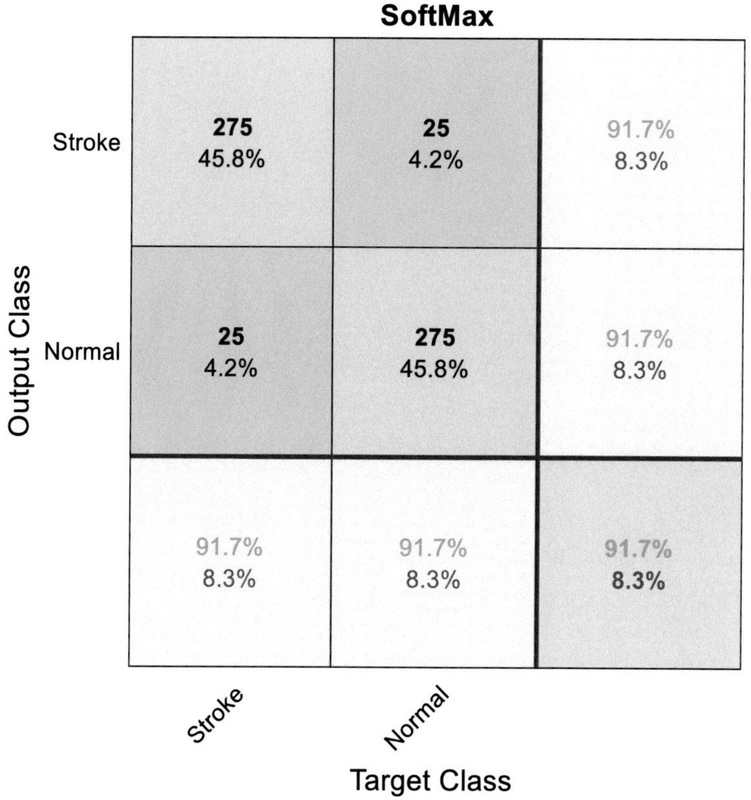

Figure 7.9 Confusion matrix achieved during the testing operation.

outcome of the SoftMax classifier and this information is then recorded separately as in Table 7.4.

A similar procedure was then repeated for other classifiers, such as NB, RF, KNN, and SVM, and these values are also recorded in Table 7.4. From this table, it can be seen that the accuracy achieved with KNN (93%) is better than that of the other classifiers, and the Glyph-plot shown in Fig. 7.12 confirms that the overall performance of the KNN is also superior to the other classifiers.

The classification of MRI is repeated with the DL + ML features using a fivefold cross-validation and the binary classifiers. During this work, improvements in the classification accuracy and overall performance are achieved for the VGG16 scheme.

Fig. 7.13 depicts the sample convergence achieved with the SVM classifier. This figure presents the essential information regarding epoch, number of iterations, error value considered, simulation time in seconds, etc.

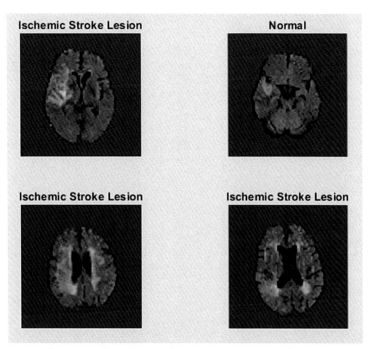

Figure 7.10 Sample classifier result achieved with SoftMax.

Table 7.3 Quality measures achieved during fivefold cross-validation with SoftMax.

Trial	TP	FN	TN	FP	ACC	PRE	SEN	SPE	NPV
1	248	52	254	46	83.6667	84.3537	82.6667	84.6667	83.0065
2	251	49	257	43	84.6667	85.3741	83.6667	85.6667	83.9869
3	264	36	270	30	89.0000	89.7959	88.0000	90.0000	88.2353
4	275	25	275	25	91.6667	91.6667	91.6667	91.6667	91.6667
5	258	42	247	53	84.1667	82.9582	86.0000	82.3333	85.4671

The results of the SVM are considered to be superior to those of the SoftMax, NB, RF, and KNN and hence the CM attained for the SVM is adopted for the assessment. The CM achieved for SVM confirms that the classification accuracy achieved is 94.5% with this data set and other related values are then recorded as in Table 7.5. Similar results were achieved for the other classifiers (Figs 7.14 and 7.15).

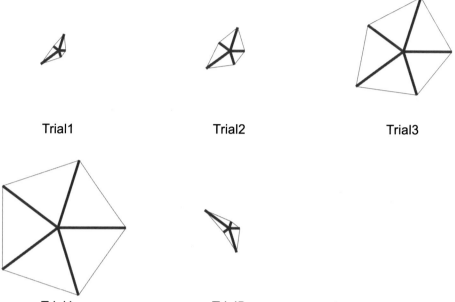

Trial1   Trial2   Trial3

Trial4   Trial5

Figure 7.11 Glyph-plot constructed using the results of five trial classifications.

Table 7.4 Quality measures attained with other binary classifiers.

Classifier	TP	FN	TN	FP	ACC	PRE	SEN	SPE	NPV
SoftMax	275	25	275	25	91.6667	91.6667	91.6667	91.6667	91.6667
NB	273	27	265	35	89.6667	88.6364	91.0000	88.3333	90.7534
RF	278	22	276	24	92.3333	92.0530	92.6667	92.0000	92.6174
KNN	280	20	278	22	93.0000	92.7152	93.3333	92.6667	93.2886
SVM	277	23	279	21	92.6667	92.9530	92.3333	93.0000	92.3841

To confirm the overall performance of the chosen classifiers, a Glyph-Plot was constructed using the Table 7.6 values and this result also confirmed that the result achieved with SVM was better than for the other classifiers. Further, this study confirms that the overall result achieved with DL + ML features offers better results with VGG16 compared to the results achieved with the DL features alone.

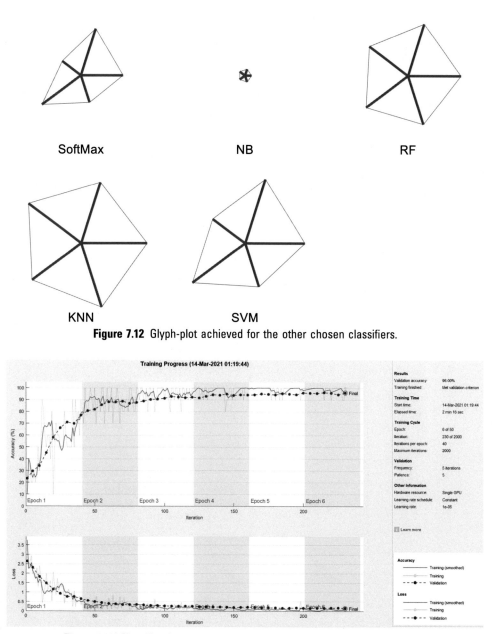

Figure 7.12 Glyph-plot achieved for the other chosen classifiers.

Figure 7.13 Classification accuracy and loss attained with DL + ML features.

From this research, it is confirmed that the ML + DL features help to achieve better IS lesion detection on the chosen MRI database. In future, the proposed method can be tested and validated on other MRI modalities, such as T1 and DW.

Table 7.5 Quality measures attained during the DL + ML feature-based classification.

Classifier	TP	FN	TN	FP	ACC	PRE	SEN	SPE	NPV
SoftMax	277	23	278	22	92.5000	92.6421	92.3333	92.6667	92.3588
NB	276	24	272	28	91.3333	90.7895	92.0000	90.6667	91.8919
RF	283	17	281	19	94.0000	93.7086	94.3333	93.6667	94.2953
KNN	284	16	280	20	94.0000	93.4211	94.6667	93.3333	94.5946
SVM	285	15	282	18	94.5000	94.0594	95.0000	94.0000	94.9495

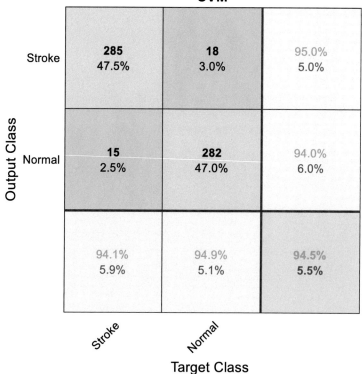

Figure 7.14 Confusion matrix achieved for SVM classifier for DL + ML features.

7.5 Discussion

The experimental outcome presented in Section 7.4 confirms that the employed DL + ML scheme helped to attain a better result for the ISLER2015 data set. This work confirmed

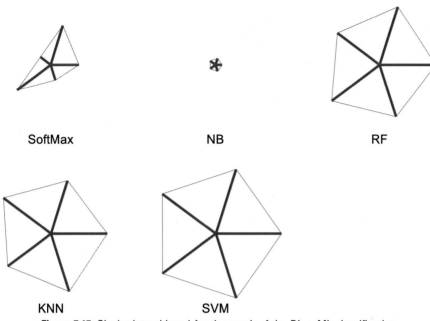

SoftMax   NB   RF

KNN   SVM

Figure 7.15 Glyph-plot achieved for the result of the DL + ML classification.

Table 7.6 Test images collected from the TCIA database.

			Images	
Image type	Dimensions	Total	Training	Testing
LGG	224 × 224 × 3	2000	1400	600
GBM	224 × 224 × 3	2000	1400	600

that the proposed scheme worked well on the flair modality MRI slices.

In order to confirm the clinical significance of the proposed scheme, the clinical-grade MRI slices of TCIA (LGG and GBM) were examined with the VGG16 with DL + ML features. Additional information about this image data set can be found in Chapter 6 (subsection 6.3.1). The task adopted in this work is classification of the MRI slices into LGG and GBM classes. Sample test images of this data set are presented in Fig. 7.16.

VGG-UNet-supported segmentation is employed to extract the tumor section from test images. The extracted images are in

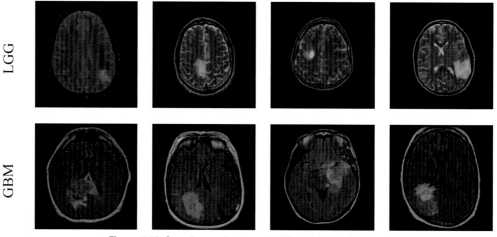

Figure 7.16 Sample test images of LGG and GBM classes.

Figure 7.17 LBP-enhanced TCIA images.

binary form, which gives the chosen ML features, like Hu moments and GLCM. To extract the LBP features, each test image is enhanced using a chosen weight (W = 1, 2, 3, and 4) and the outcome of this task is depicted in Fig. 7.17; from this figure, the LBP features are extracted.

The VGG16-supported MRI classification task is then implemented on the considered data to extract the DL features and the various intermediate images collected from the VGG16 are depicted in Fig. 7.18.

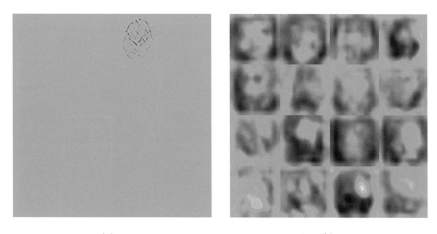

(a)   (b)

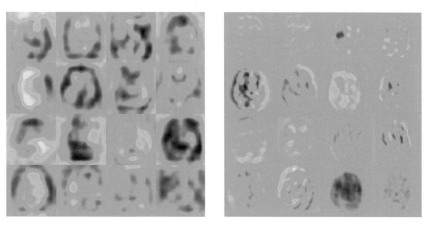

(c)   (d)

Figure 7.18 Image outcomes from various convolutional layers: (A) ConvNet1; (B) ConvNet2; (C) ConvNet3; (D) ConvNet4.

Figs. 7.18A−D depict the processed images of CovNet1 to Cov-Net4, respectively. The outcome of the CovNet5 will be the features and it is then transferred to the fully connected layers. After possible dropout, the DL features are then combined with the ML features and the classification task is performed. Initially, the classification is performed using SoftMax using a fivefold cross-validation and the achieved results are presented with the best result chosen as the final result by SoftMax. Figs. 7.19 and 7.20 present the convergence of the accuracy and loss for the trial 2 operation, which confirms that the classification accuracy obtained for this data set is >90%.

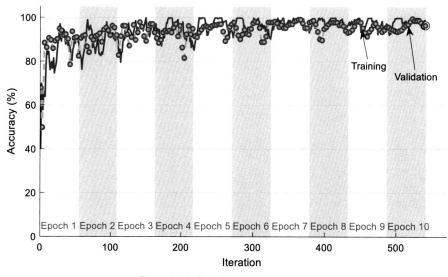

Figure 7.19 Classification accuracy.

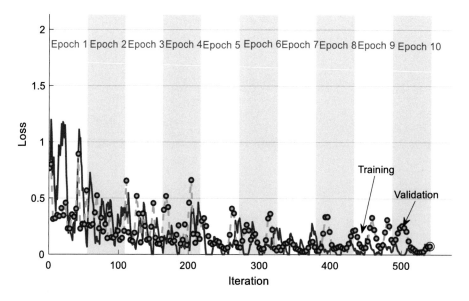

Figure 7.20 Classification loss.

The corresponding CM is presented in Fig. 7.21, which confirms the attained QM during the trial 2 classification.

All the computed QM are presented in Table 7.7, with this information proving that the projected work helped to get improved

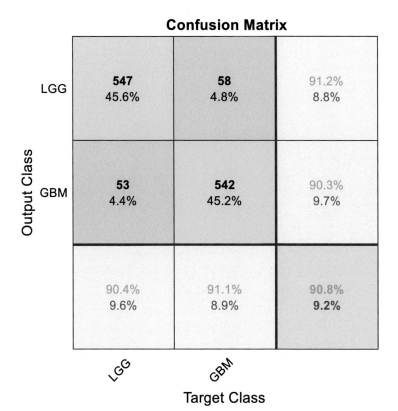

Figure 7.21 Confusion matrix obtained for trial 2 with SoftMax.

Table 7.7 Classifier performance during fivefold cross-validation.

Classification	TP	FN	TN	FP	AC	PR	SE	SP	NPV
Trial1	544	56	543	57	90.58	90.52	90.67	90.50	90.65
Trial2	547	53	542	58	**90.75**	90.41	**91.17**	90.33	**91.09**
Trial3	541	59	545	55	90.50	**90.77**	90.17	**90.83**	90.23
Trial4	538	62	543	57	90.08	90.42	89.67	90.50	89.75
Trial5	536	64	539	61	89.58	89.78	89.33	89.83	89.39

Bold represents the best values.

values of AC, SE, and NPV during trial 2 and better PR and SP during trial 3. To confirm the overall performance, each result is individually compared, as depicted in Fig. 7.22, which confirms that trial 2 provides the overall best result as compared with trial 3.

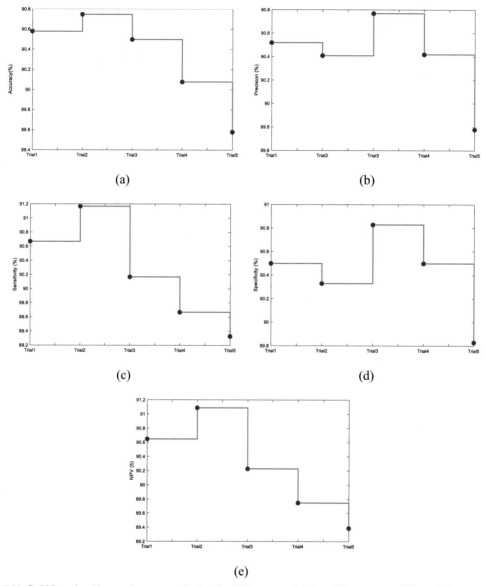

Figure 7.22 SoftMax classifier performances during fivefold cross-validation: (A) accuracy; (B) precision; (C) sensitivity; (D) specificity (E) NPV.

The proposed classification task is then repeated using other classifiers like, NB, RF, KNN, and SVM, with these outcomes depicted in Table 7.8. The tumor detection performance of the VGG16 is also authorized using other pretrained schemes, such as VGG19, AlexNet, resNet18, and ResNet50, and these results are also included in Table 7.8.

Table 7.8 Performance evaluation of VGG16 with other pretrained classifiers.

Scheme	Classifier	TP	FN	TN	FP	ACC	PRE	SEN	SPE	NPV
VGG16	SoftMax	547	53	542	58	90.75	90.41	91.17	90.33	91.09
	NB	539	61	640	60	89.92	89.98	89.83	90.00	89.85
	RF	537	63	538	62	89.58	89.64	89.50	89.67	89.52
	KNN	548	52	544	56	91.00	90.73	91.33	90.67	91.28
	SVM	548	52	541	59	90.75	90.28	91.33	90.17	91.23
VGG19	SoftMax	538	62	537	63	89.58	89.52	89.67	89.50	89.65
	NB	533	67	531	69	88.67	88.54	88.83	88.50	88.80
	RF	528	72	533	67	88.42	88.74	88.00	88.83	88.10
	KNN	526	74	530	70	88.00	88.26	87.67	88.33	87.75
	SVM	541	59	545	55	90.50	90.77	90.17	90.83	90.23
AlexNet	SoftMax	554	46	565	35	93.25	94.06	92.33	94.17	92.47
	NB	550	50	558	42	92.33	92.91	91.67	93.00	91.78
	RF	563	37	558	42	93.42	93.06	93.83	93.00	93.78
	KNN	557	43	556	44	92.75	92.68	92.83	92.67	92.82
	SVM	552	48	551	49	91.92	91.85	92.00	91.83	91.99
ResNet18	SoftMax	552	48	561	39	92.75	93.40	92.00	93.50	92.12
	NB	548	52	563	37	92.58	93.68	91.33	93.83	91.54
	RF	553	47	551	49	92.00	91.86	92.17	91.83	92.14
	KNN	550	50	549	51	91.58	91.51	91.67	91.50	91.65
	SVM	559	41	560	40	93.25	93.32	93.17	93.33	93.18
ResNet50	SoftMax	582	18	578	22	96.67	96.36	97.00	96.33	96.98
	NB	577	23	571	29	95.67	95.21	96.17	95.17	96.13
	RF	573	27	583	17	96.33	97.12	95.50	97.17	95.57
	KNN	584	16	576	24	96.67	96.05	97.33	96.00	97.30
	SVM	583	17	581	19	97.00	96.84	97.17	96.83	97.16

These outcomes verify that the proposed VGG16 is efficient in detecting abnormalities in the ISLES2015 and TCIA data sets. This also confirms that this scheme works well for flair as well as T2 modality MRI slices.

7.6 Conclusion

Ischemic stroke (IS) is a common abnormality that arises in the brain due to a reduction in the oxygen supply, and therefore timely detection and treatment are necessary to treat the patient. MRI-supported IS lesion detection is widely adopted for identifying the location and severity of IS lesions, and hence a number of methods have been suggested by researchers for automated diagnosis. The proposed work implements the VGG16-supported scheme to detect disease with better accuracy. This procedure employs the VGG-UNet-based segmentation procedure to extract the IS lesion area from the flair modality MRI and then the necessary ML features are extracted. These ML features are then linked with the DL features of VGG16 and the disease detection task is then implemented with DL and DL + ML features. In this work, a binary categorization is implemented and the proposed work with DL + ML helps to accomplish an improved product on the preferred images of the ISLES2015 database. A similar practice is repeated with test images from the TCIA data set and a better result is achieved with DL + ML features. In the future, clinically collected MRI slices could be tested with the proposed scheme.

References

[1] Chou WY, Tien PT, Lin FY, Chiu PC. Application of visually based, computerised diagnostic decision support system in dermatological medical education: a pilot study. Postgraduate Medical Journal 2017; 93(1099):256−9.

[2] Guchelaar HJ, Kalmeijer MD. The potential role of computerisation and information technology in improving prescribing in hospitals. Pharmacy World and Science 2003;25(3):83−7.

[3] Grøvik E, Yi D, Iv M, Tong E, Rubin D, Zaharchuk G. Deep learning enables automatic detection and segmentation of brain metastases on multisequence MRI. Journal of Magnetic Resonance Imaging 2020;51(1):175−82.

[4] Zhou Z, Sanders JW, Johnson JM, Gule-Monroe MK, Chen MM, Briere TM, et al. Computer-aided detection of brain metastases in T1-weighted MRI for stereotactic radiosurgery using deep learning single-shot detectors. Radiology 2020;295(2):407−15.

[5] Altinkaya E, Polat K, Barakli B. Detection of Alzheimer's disease and dementia states based on deep learning from MRI images: a comprehensive review. Journal of the Institute of Electronics and Computer 2020;1(1): 39−53.

[6] Gautam A, Raman B. Towards effective classification of brain hemorrhagic and ischemic stroke using CNN. Biomedical Signal Processing and Control 2021;63:102178.

[7] Subudhi A, Dash M, Sabut S. Automated segmentation and classification of brain stroke using expectation-maximization and random forest classifier. Biocybernetics and Biomedical Engineering 2020;40(1):277−89.

[8] Kadry S, Damaševičius R, Taniar D, Rajinikanth V, Lawal IA. U-net supported segmentation of ischemic-stroke-lesion from brain MRI slices. In: 2021 seventh international conference on bio signals, images, and instrumentation (ICBSII). IEEE; March 2021. p. 1−5.

[9] Hemanth DJ, Rajinikanth V, Rao VS, Mishra S, Hannon NM, Vijayarajan R, Arunmozhi S. Image fusion practice to improve the ischemic-stroke-lesion detection for efficient clinical decision making. Evolutionary Intelligence 2021:1−11.

[10] Lin D, Rajinikanth V, Lin H. Hybrid image processing-based examination of 2D brain MRI slices to detect brain tumor/stroke section: a study. In: Signal and image processing techniques for the development of intelligent healthcare systems. Singapore: Springer; 2021. p. 29–49.

[11] Johnson W, Onuma O, Owolabi M, Sachdev S. Stroke: a global response is needed. Bulletin of the World Health Organization 2016;94(9):634.

[12] Maier O, Wilms M, von der Gablentz J, Krämer U, Handels H. March). Ischemic stroke lesion segmentation in multi-spectral MR images with support vector machine classifiers. In: Medical imaging 2014: computer-aided diagnosis, vol. 9035. International Society for Optics and Photonics; 2014. p. 903504.

[13] Maier O, Schröder C, Forkert ND, Martinetz T, Handels H. Classifiers for ischemic stroke lesion segmentation: a comparison study. PLoS One 2015; 10(12):e0145118.

[14] Maier O, Wilms M, von der Gablentz J, Krämer UM, Münte TF, Handels H. Extra tree forests for sub-acute ischemic stroke lesion segmentation in MR sequences. Journal of Neuroscience Methods 2015;240:89–100.

[15] Maier O, Menze BH, von der Gablentz J, Häni L, Heinrich MP, Liebrand M, et al. ISLES 2015-A public evaluation benchmark for ischemic stroke lesion segmentation from multispectral MRI. Medical Image Analysis 2017;35: 250–69.

[16] Subbanna NK, Rajashekar D, Cheng B, Thomalla G, Fiehler J, Arbel T, Forkert ND. Stroke lesion segmentation in FLAIR MRI datasets using customized Markov random fields. Frontiers in Neurology 2019;10:541.

[17] Zhang L, Song R, Wang Y, Zhu C, Liu J, Yang J, Liu L. Ischemic stroke lesion segmentation using multi-plane information fusion. IEEE Access 2020;8: 45715–25.

[18] Kumar A, Upadhyay N, Ghosal P, Chowdhury T, Das D, Mukherjee A, Nandi D. CSNet: a new DeepNet framework for ischemic stroke lesion segmentation. Computer Methods and Programs in Biomedicine 2020;193: 105524.

[19] Rajinikanth V, Satapathy SC. Segmentation of ischemic stroke lesion in brain MRI based on social group optimization and Fuzzy-Tsallis entropy. Arabian Journal for Science and Engineering 2018;43(8):4365–78.

[20] Zhang Y, Liu S, Li C, Wang J. Application of deep learning method on ischemic stroke lesion segmentation. Journal of Shanghai Jiaotong University (Science) 2021:1–13.

[21] Simonyan K, Zisserman A. Very deep convolutional networks for large-scale image recognition. arXiv; 2014. preprint arXiv:1409.1556.

[22] Dey N, Zhang YD, Rajinikanth V, Pugalenthi R, Raja NSM. Customized VGG19 architecture for pneumonia detection in chest X-rays. Pattern Recognition Letters 2021;143:67–74.

[23] Ahuja S, Panigrahi BK, Dey N, Rajinikanth V, Gandhi TK. Deep transfer learning-based automated detection of COVID-19 from lung CT scan slices. Applied Intelligence 2021;51(1):571–85.

[24] Dey N, Rajinikanth V, Fong SJ, Kaiser MS, Mahmud M. Social group optimization–assisted Kapur's entropy and morphological segmentation for automated detection of COVID-19 infection from computed tomography images. Cognitive Computation 2020;12(5):1011–23.

[25] Tang KJW, Ang CKE, Theodoros C, Rajinikanth V, Acharya UR, Cheong KH. Artificial intelligence and machine learning in emergency medicine. Biocybernetics and Biomedical Engineering; 2020.

[26] Pugalenthi R, Rajakumar MP, Ramya J, Rajinikanth V. Evaluation and classification of the brain tumor MRI using machine learning technique. Journal of Control Engineering and Applied Informatics 2019;21(4):12–21.

[27] Rajinikanth V, Kadry S. Development of a framework for preserving the disease-evidence-information to support efficient disease diagnosis. International Journal of Data Warehousing and Mining 2021;17(2):63–84.

[28] Gudigar A, Raghavendra U, Devasia T, Nayak K, Danish SM, Kamath G, et al. Global weighted LBP based entropy features for the assessment of pulmonary hypertension. Pattern Recognition Letters 2019;125:35–41.

[29] Rajinikanth V, Joseph Raj AN, Thanaraj KP, Naik GR. A customized VGG19 network with concatenation of deep and handcrafted features for brain tumor detection. Applied Sciences 2020;10(10):3429.

[30] Bhandary A, Prabhu GA, Rajinikanth V, Thanaraj KP, Satapathy SC, Robbins DE, et al. Deep-learning framework to detect lung abnormality—A study with chest X-Ray and lung CT scan images. Pattern Recognition Letters 2020;129:271–8.

Index

'*Note:* Page numbers followed by "f" indicate figures and "t" indicate tables.'

Printed in the United States
by Baker & Taylor Publisher Services